Deluge

The Bible was right!!!

Ivo Jara Gimeno

To my creationist friend

Jeremy S.

Thanks for helping me find information that is not mainstream, resources that do not deny what is described by hundreds of cultures from the past.

Ivo

Proofreading by: Paola Antivilo Mattman

Did the deluge happen?

Mainstream science has assured everybody that "The great flood" didn´t happen, that there is no evidence of it. We can see an example of this below.

"...there seems to be a good reason for thinking that some and probably many diluvial traditions are merely exaggerated reports of floods which actually occurred, whether the result of heavy rain, earthquake-waves (Tsunamis), or other causes. All such traditions therefore, are partly legendary and partly mythical: so far as they preserve reminiscences of floods which really happened, they are legendary; so far as they describe universal deluges which never happened, they are mythical."
James Frazer, Folklore in the Old Testament (1918)

This has not changed in recent years; for example, there is a paper published: ***Twenty-one Reasons Noah's Worldwide Flood Never Happened***, published by Lorence G. Collins of California State University Northridge, and many more like this one, another example is Wikipedia, which says *"Flood geology (also creation geology or diluvial geology) is the attempt to interpret and reconcile geological features of the Earth in accordance with a literal belief in the global flood described in Genesis 6–8."*, As you can see they write *"...the attempt"*. A global flood just rubs a lot of people the wrong way, and they will repeat over and over again that *"There is no recorded flood at the time of Noah."* But there is; it is recorded in The Bible. The purpose of this book is to use scientific evidence to prove that the deluge or Noah's flood actually happened, how it happened and exactly when. The "myth" of the Deluge, the great flood or however you want to call it, as documented in the Bible, is also documented in the oral and written traditions of many cultures in history and pre-history of mankind.

When you begin researching a subject, you have an idea of where it's going to take you; let's call it a thesis, and my thesis or idea was that the deluge happened and that it was such a catastrophic event that most cultures and religions recorded it as "the death" of mankind by God or "the Gods" and a new beginning. Many have thought about this, and some have even dared to say it. Believing the Deluge happened is not news, although mainstream science is still at odds with it, but trying to identify when and how it actually happened is another story.

I was raised catholic, and whenever I read the Old Testament stories I would be impressed by the descriptions of the wrath of God, but to a kid it was essentially a 40 day and 40 night storm. In time I began reading the scriptures in detail and although I became separated from religion, I began reading adult versions of the scriptures; it is in these versions for adults, not in the children's books; that I learned of water gushing out of the ground, winds, massive wildfires and fire in the sky.

Now, as an adult I began reading in depth about it, scientific papers, mainstream and fringe authors, and decided to dig deep into the subject. Let me give you an example, a tree falls in your neighbor's yard while you are at work, you come back home and the sidewalk is covered in leaves, the tree is not there; you never paid much attention to your neighbor's yard, you know there are trees there, but you are not really sure if your neighbor pruned his trees, and that is where all the leaves came from, or that something else happened, then you see a tree stump, you walk into your house and your son says "Dad, there was a massive crash outside, then a truck came and I heard chainsaws". You walk across the street and you other neighbor tells you about the whole affair across the street, you ask "What happened?", he answers back "I don't know, but a truck came and took big logs and a bunch of branches", then you walk across the street and ask you next door neighbor and he tells you that his tallest tree fell because the soil under it failed. So now, all the other stories you got first make sense, this is the case that I am making, if you get all the stories of an event, eventually everything will fall into place.

So my "thesis" was that there was one great flood and that it must have been recorded, if it really happened there must also be geological and scientific evidence of such an event. If we are told by scientists it didn't happen, and the Bible says it did, somebody else must have seen something, if it didn't happen it should make sense, it's not like a global flood can be just ignored or happened unnoticed by the whole globe. So the search began.

I began with the bible, after all it is the text I was familiar with, I wanted to see exactly which mentions of the flood were actually there and to what level of detail it described it. If the deluge was as described in the Bible it must certainly have been recorded by someone else, not just by those who wrote the Old Testament; and if this was the case, I should find all these records and analyze them to find the actual event and its causes. For Christians the causes are explained in the Bible, mankind became wicked, corrupt and deserved to die, and for no clear reason, all the rest of the animals as well... The Earth had to be cleansed of evil, sin and degradation and God did it with water, by drowning every living thing on the planet with the exception of those He deemed worthy of saving. Christians would be surprised at how many hundreds of cultures, their religions, myths and traditions actually confirm this. While searching for anything available to read on the matter, I became aware of the book of Enoch, which was not included in the Bible, I also read it and did find some more specific reasons as to why God had decided to flood the Earth and kill every living thing; I will not discuss the content now, but it is a very interesting read, specially for Christians.

Genesis 6 tells us:

In the six hundredth year of Noah's life, on the seventeenth day of the second month—on that day all the springs of the great deep burst forth, and the floodgates of the heavens were opened. And rain fell on the earth forty days and forty nights.

For forty days the flood kept coming on the earth, and as the waters increased they lifted the ark high above the earth.

The waters rose and increased greatly on the earth, and the ark floated on the surface of the water.

They rose greatly on the earth, and all the high mountains under the entire heavens were covered.

The waters rose and covered the mountains to a depth of more than fifteen cubits.

Every living thing that moved on land perished—birds, livestock, wild animals, all the creatures that swarm over the earth, and all mankind.

Everything on dry land that had the breath of life in its nostrils died.

Every living thing on the face of the earth was killed; people and animals and the creatures that move along the ground and the birds were wiped from the earth. Only Noah was left and those with him in the ark...

By the first day of the first month of Noah's six hundred and first year, the water had dried up from the earth. Noah then removed the covering from the ark and saw that the surface of the ground was dry.

By the twenty-seventh day of the second month the earth was completely dry.

A week later Noah sent out a dove, and on its third flight it returned with an olive leaf plucked from the Mount of Olives in Jerusalem, for the Holy Land had not suffered from the flood. Noah wept at the devastation when he left the ark, and Shem offered a thank-offering;

Please keep in mind that the Hebrews do have a flood, but it is not planetary, because they believe that *"for the Holy Land had not suffered from the flood"* So, this sole statement provides us with an important clue, according to the Hebrews, Noah could not have lived in the holy land, a fact that has been argued by some scholars who place the building of the ark somewhere around the fertile crescent of the Tigris and Euphrates a.k.a. Mesopotamia.

It was there where the earliest Flood and Ark story has been found, this story is in tablet XI of XII of the Epic of Gilgamesh, these tablets are written in Akkadian, tablet XI or The Flood Tablet, is part of a series that tells the Gilgamesh epic, from Nineveh, 7th century BC; this tablet is currently on display in the British Museum in London.

In the Epic of Gilgamesh, the story says (summarized): *Utnapishtim was granted immortality after building a ship called **Preserver of Life** and using it to survive the "great flood.", Utnapishtim built his ship according to the instructions of the God Enki, and when the water came, he sailed for 19 days and then released a raven, the raven saw that the water receded and did not return. Utnapishtim then set all the animals free, and then made a sacrifice to the gods. Utnapishtim sailed with all of his relatives and all species of animals aboard his ship to save mankind.*

The Flood myths

Now, a great flood very far in the past is a story that has been told countless times.

Below you will find a list of civilizations, cultures and tribes which have a global flood myth.

Europe

Greek, Arcadian, Samothrace, Roman, Scandinavian, German, Celtic, Welsh, Lithuanian, Transylvanian, Gypsy, Russian

Middle East

Sumer, Egypt, Babylon, Assyrian, Chaldean, Hebrew, Islamic, Turkey, Persian, Zoroastrian

Africa

Cameroon, Masai (East Africa), Komililo Nandi, Kwaya (Lake Victoria), Southwest Tanzania, Pygmy, Ababua (northern Zaire), Kikuyu (Kenya), Bakongo (west Zaire), Bachokwe? (southern Zaire), Lower Congo, Basonge, Bena-Lulua (Congo River, southeast Zaire), Yoruba (southwest Nigeria), Efik-Ibibio (Nigeria), Ekoi (Nigeria), Mandingo (Ivory Coast)

Asia

Vogul, Samoyed (north Siberia), Yenisey-Ostyak (north central Siberia), Kamchadale (northeast Siberia), Altaic (central Asia), Tuvinian (Soyot) (north of Mongolia), Mongolia, Buryat (eastern Siberia), Sagaiye (eastern Siberia), Hindu, Bhil (central India), Kamar (Raipur District, Central India), Assam, Tamil (southern India), Lepcha (Sikkim), Tibet, Singpho (Assam), Lushai (Assam),

Lisu (northwest Yunnan, China), Lolo (southwestern China), Jino (southern Yunnan, China), Karen (Burma), Chingpaw (Upper Burma), China, Korea, Munda (north-central India), Santal (Bengal), Ho (southwestern Bengal), Bahnar (Cochin China), Kammu (northern Thailand), Andaman Islands (Bay of Bengal), Zhuang (China), Sui (southern Guizhou, China), Shan (Burma), Tsuwo (Formosa interior), Bunun (Formosa interior), Ami (eastern Taiwan), Benua-Jakun (Malay Peninsula), Kelantan (Malay Peninsula), Ifugao (Philippines), Kiangan Ifugao, Atá (Philippines), Mandaya (Philippines), Tinguian (Luzon, Philippines), Batak (Sumatra), Nias (an island west of Sumatra), Engano (another island west of Sumatra), Dusun (British North Borneo), Dyak (Borneo), Ot-Danom (Dutch Borneo), Toradja (central Celebes), Alfoor (between Celebes and New Guinea), Rotti (southwest of Timor), Nage (Flores)

Australia

Arnhem Land (northern Northern Territory), Maung (Goulburn Islands, Arnhem Land), Gunwinggu (northern Arnhem Land), Gumaidj (Arnhem Land), Manger (Arnhem Land), Fitzroy River area (Western Australia), Australian, Mount Elliot (coastal Queensland), Western Australia, Andingari (South Australia), Wiranggu (South Australia), Narrinyeri (South Australia), Victoria, Lake Tyres (Victoria), Kurnai (Gippsland, Victoria), southeast Australian, Maori (New Zealand)

Pacific Islands

Kabadi (New Guinea), Valman (northern New Guinea), Mamberao River (Irian Jaya), Samo-Kubo (western Papua New Guinea), Papua New Guinea, Palau Islands (Micronesia), western Carolines, New Hebrides, Lifou (one of the Loyalty Islands), Fiji, Samoa, Nanumanga (Tuvalu, South Pacific), Mangaia (Cook Islands), Rakaanga (Cook Islands), Raiatea (Leeward Group, French Polynesia), Tahiti, Hawaii

North America

Innuit, Eskimo (Orowignarak, Alaska), Norton Sound Eskimo, Central Eskimo, Tchiglit Eskimo (Arctic Ocean), Herschel Island Eskimo, Netsilik Eskimo, Greenlander, Tlingit (southern Alaska coast), Hareskin (Alaska), Tinneh (Alaska and south), Loucheux (Dindjie) (Alaska), Dogrib and Slave (Tinneh tribes), Kaska (northern inland British Columbia), Thompson Indians (British Columbia), Sarcee (Alberta), Tsetsaut, Haida (Queen Charlotte Is., British Columbia), Tsimshian (British Columbia), Kwakiutl (British Columbia), Kootenay (southeast British Columbia), Squamish (British Columbia), Bella Coola (British Columbia), Lillooet (Green River, British Columbia), Makah (Cape Flattery, Washington), Klallam (northwest Washington), Skokomish (Washington), Skagit (Washington), Quillayute (Washington), Nisqually (Washington), Twana (Puget Sound, Washington), Kathlamet, Cascade Mountains, Spokana, Nez Perce, Cayuse (eastern Washington), Yakima (Washington), Warm Springs (Oregon), Joshua (southern Oregon), Smith River (northern California coast), Wintu (north central California), Maidu (central California), Northern Miwok (central California), Tuleyome Miwok (near Clear Lake, California), Olamentko Miwok (Bodega Bay, California) Ohlone (San Francisco to Monterey, California), Kato (Mendocino County, California), Shasta (northern California interior), Pomo (north central California), Salinan (California), Yuma (western Arizona, southern California), Havasupai (lower Colorado River), Ashochimi (California), Yurok (north California coast), Blackfoot (Alberta and Montana), Cree (Canada), Timagami Ojibway (Canada), Chippewa (Ontario, Minnesota, Wisconsin), Ottawa, Menomini (Wisconsin-Michigan border), Cheyenne (Minnesota), Yellowstone, Montagnais (northern Gulf of St. Lawrence), Micmac (eastern Maritime Canada), Algonquin (upper Ottowa River), Lenape (Delaware) (Delaware to New York), Cherokee (Great Lakes area; eastern Tennessee), Mandan

(North Dakota), Lakota, Choctaw (Mississippi), Natchez (Lower Mississippi), Chitimacha (Southern Louisiana), Caddo (Oklahoma, Arkansas), Pawnee (Nebraska), Navajo (Four Corners area), Jicarilla Apache (northeastern New Mexico), Sia (northeast Arizona), Acagchemem (near San Juan Capistrano, California), Luiseño (Southern California), Pima (southwest Arizona), Papago (Arizona), Hopi (northeast Arizona), Zuni (New Mexico),

Central America

Tarascan (northern Michoacan, Mexico), Michoacan (Mexico), Yaqui (Sonoran, Northern Mexico), Tarahumara (Northern Mexico), Huichol (western Mexico), Cora (east of the Huichols), Tepecano (southeast of the Huichols), Tepehua (eastern Mexico), Toltec (Mexico), Nahua (central Mexico), Tlaxcalan (central Mexico), Tlapanec (south central Mexico), Mixtec (northern Oaxaca, Mexico), Zapotec (Oaxaca, southern Mexico), Trique (Oaxaca, southern Mexico), Totonac (eastern Mexico), Chol (southern Mexico), Tzeltal (Chiapas, southern Mexico), Quiché (Guatemala), Maya (southern Mexico and Guatemala), Popoluca (Veracruz, Mexico), Nicaragua, Panama, Carib (Antilles)

South America

Acawai (Orinoco), Arekuna (Guyana), Makiritare (Venezuela), Macusi (British Guyana), Muysca (Colombia), Yaruro (southern Venezuela), Yanomamö (southern Venezuela), Tamanaque (Orinoco), Arawak (Guyana), Pamary, Abedery, and Kataushy (Purus R., Brazil), Ipurina (Upper Amazon), Jivaro (eastern Ecuador), Shuar (Andes), Murato (eastern Ecuador), Cañari (Quito, Ecuador), Guanca and Chiquito (Peru), Ancasmarca (near Cuzco, Peru), Canelos Quechua, Quechua, Inca (Peru), Colla (high Andes), Chiriguano (southeast Bolivia), Chorote (Eastern Paraguay), Eastern Brazil (Rio de Janiero region), Eastern Brazil (Cape Frio region), Caraya (Araguaia River, central Brazil), Coroado (south Brazil), Araucania (coastal Chile), Toba (northern

Argentina), Selk'nam (southern tip of Argentina), Yamana (Tierra del Fuego)

This list was compiled by Mark Isaak (or at least, this is where I found it): http://www.talkorigins.org/faqs/flood-myths.html His actual list includes a brief (and sometimes not so brief) summary of each one of the myths.

Now, let's think seriously about this, the flood story, told by cultures all around the planet, from the land of the Inuit to the Australian aborigines; Even the Maori of New Zealand and the Rapa Nui of Easter Island tell the story… For centuries scientists have denied the deluge as religious nonsense, but, can all these cultures be wrong, did all of them create the same fictional story? This doesn't make sense.

So, just as an exercise and using an example which is very far away from the Middle East and the Biblical story, now let's take a look at the Aztec deluge myth.

It was Quetzalcoatl, the feathered serpent who created the first humans; he used grey ash to create us. In the first time, the Earth gave its riches for free to the new men, they had easy lives, without pain or suffering, after some time they became greedy; they did not honor their creator anymore and took the riches for granted.

Quetzalcoatl was furious and decided to put an end to them with a great flood. However, there was one couple that was not greedy, their names were Teta and his wife Nena, Quetzalcoatl decided to spare them and gave them instructions.

"Find the tallest and strongest hollow tree, hide in it and you shall live, take only one ear of corn with you and no more", Teta and Nena did what they were told...

Quetzalcoatl sent water crashing on his creation, the greedy men and women turned into fish, and are still fish today; Teta and Nena survived as Quetzalcoatl promised, they repopulated the earth with a new race, humble and always aware of the Gods who created them. However Coatlicue, Mother Earth, was never as generous with its riches as in the past.

Please notice that Quetzalcoatl *"Sent water crashing on his creation..."* No rain here.

We could spend months reading flood stories from different parts of the world and they will all tell a similar general story:

1) God or the Gods created mankind

2) Mankind becomes impure, or greedy, or defiant or wicked in some way or another.
3) God or the Gods decide to kill them all and start over
4) There is a man or a couple who are not wicked and God or the Gods decide to spare him and his family, sometimes with and sometimes without animals.
5) God sends a messenger to warn them, warns them himself or the man or couple who are not wicked dream about the warning.
6) The great flood comes and everything is killed.
7) The waters recede or the land is forever flooded.
8) The couple repopulates the Earth with good or clean people once the water recedes or they move to another place because the original one was flooded and they could not return.

The world was destroyed, the only thing powerful enough to do this was God or "The Gods", and in order to survive, humans had to obey him or them and "behave" in the future.

Were humans wicked people and got punished? Maybe, although it is not my place to judge, of course if a disaster is too big and harmful, the only explanation is a deity. With this last paragraph, I am not denying that there was a deluge or that a God caused it, I simply don't know; after all, I was not there, and the people who told the story were.

The Aztecs and many other cultures found an explanation; and just like Judeo Christian and Abrahamic faiths, developed a set of rules to please God and avert another punishment.

The Aztecs ended their flood myth with:

...Coatlicue (mother earth) became known as "The grandmother", she wore a skirt made out of snakes and a necklace made from human hands and hearts adorned with a skull. Her hands and feet were decorated with animal claws. She became hungry for human blood and demanded periodic human sacrifices and humans had to give her one living human heart each year, and if they didn't Coatlicue would bring starvation to Earth.

That "one human heart every year" thing was not the nicest of rules, but cultures sometimes have different standards. We also have to keep in mind that these societies were not like ours, and sacrifice may mean a lamb for one society, or a human to another. Another difference is that many of these people usually conveyed information by handing it down using oral tradition and, as an example; the oldest text of the bible was not written until very long after the flood described in it actually happened. In the case of the Bible, God displayed a rainbow, which symbolized his promise of never flooding the Earth again, but of course all contracts come with fine print, and God was very specific, he said he wouldn't flood the Earth ever again, but what he didn't say is that he wouldn't destroy the planet again; but that is another story covered in Revelation.

A text of the flood discovered in the charred Ein Gedi scroll is 100 percent identical to the version of the Book of Leviticus that has been in use for centuries, scholar Emmanuel Tov from the Hebrew University of Jerusalem said: *"This text has not changed in 2,000 years."*

Now, if you are a creationist Christian, you would date the deluge about 1,656 years after creation and if you have the age of the planet as roughly 6,000 years, the first written bible we know of would come to be about 2,300 years after the deluge, which is a long time for something that was not written, however this does not make it necessarily inaccurate.

If you are a creationist Christian, specifically a young Earth creationist, you will be surprised with what I am about to tell you, but here it comes. There are millions of people in the planet that do not have a clue of what you believe; most people think that the history of our planet is always the same and that if you are a creationist, you believe the world was created and that everything else remains the same, of course there are people who believe that, that the world was created by God, but that this happened 4.3 billion years ago, so it 's essentially the scientific view but with God added as a kickstarter. So, please allow me to provide an explanation of what young earth creationists believe, so those billions of people in the world can be aware of what we are talking about.

What is a young Earth creationist?

A young Earth Creationist is a person who believes the Bible is a correct account of the events that led to the existence of the world as we know it, so to them, Genesis is a literal account of what happened. This means the planet is 6,000 years old, and all living things were created by God using His supernatural abilities, young Earth creationists allow for evolution within that timeframe and can provide you with examples of this. The age of the Earth is calculated using the Bible, so to them, Adam, the first man and Eve, the first woman were created by God and Adam is said to have lived 930 years, FYI Adam and Eve had 56 Children.

Now, the Bible does not say when Cain and Abel were born, but it does say when their younger brother Seth was born, according to Genesis 5:3 Seth was born when Adam was 130 years old and in turn Seth lived to be 912 years old. So what young Earth creationism does to calculate the age of the planet is add up all the ages of these people, their children, and their grandchildren and so on until the time of Christ. This gives an age for the planet of about 6,000 years

As I just said, if you are not a creationist, you will be surprised of the age young Earth creationists attribute to the Earth – Universe. It can be broken down as follows:

First six days... Because (Genesis 1:1–2:4), God on the sixth day of Creation created all the living creatures and, "in his own image," man, both "male and female." God then blessed the couple, told them to be "fruitful and multiply," then:

Adam To Abraham =	About 2,000 years
Abraham to Jesus =	About 2,000 years
Jesus to present =	About 2,000 years
Total =	**About 6,000 years**

If you are not a creationist, for you the Earth is about 4,3 billion years old, and humans have been on earth for about 300,000 to 500,000 years, this "humans" means with a modern size brain, and the period of time we will be covering in this book is about the last 15,000 years, but actually we will focus in detail on the last 6,000 years, yes; we will deal with the same time period as the creationist timeline. Why would I limit the discussion to this short time period? Because this book deals specifically with Noah's flood, if it happened or not, when did it happen and what actually caused it, and for a creationist anything I mention from before 6,000 years ago is irrelevant anyway. So, for all (young earth) creationists who are reading this book, whenever I mention times before 6,000 years ago, it is not for you, it is for the non creationists. Will that deny God caused the deluge? No, will it change the date of the deluge? No. I don't need to, I am writing these lines to find a scientific explanation or account of the deluge outside the one in The Bible, prove once and for all that it did happen and how and be done with it. I am not here to discuss creation, the big bang, or if matter can be generated form nothing. If at all I will describe the mechanism used by nature, or God, whichever you prefer to cause this event. I will mention times before 6,000 years, this book should not become a discussion on creation or dates, but those who accept the scientific timeline of Earth need to be able to place these events in their timeline, which happens to contain (or include) the creationist timeline.

The Timelines

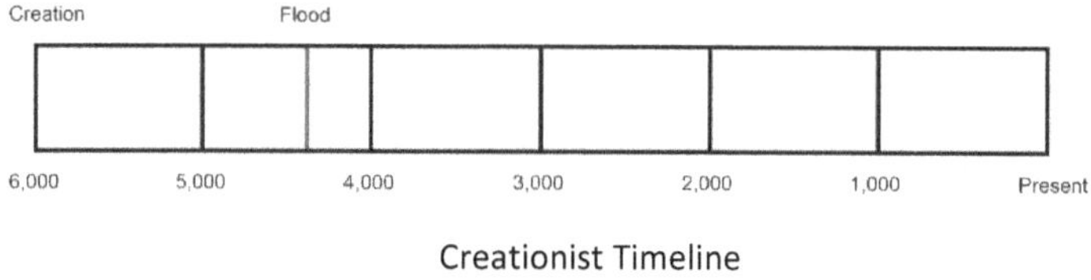

Creationist Timeline

Scientific Timeline

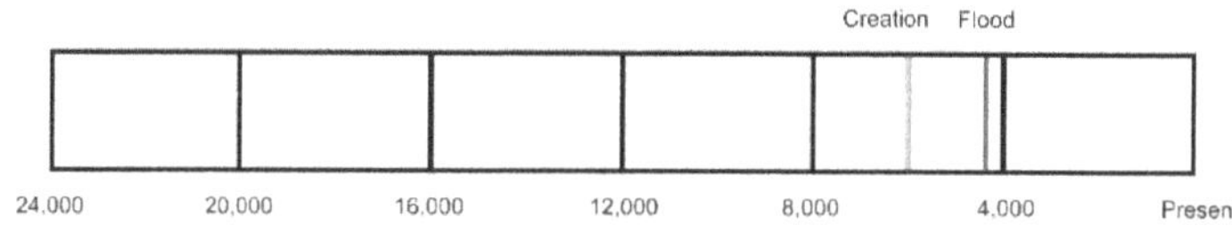

Scientific Timeline, last 24,000 years

The graphs above are to provide you with a visual representation of what we are talking about.

Before we continue with the Bible, which became known thousands of years after these events happened, let's talk a little bit about how oral tradition actually works.

Oral tradition is usually not treated very well by anthropologists, historians and scientists in general, every time a historian wants to discredit or cast doubt on oral tradition they use the "telephone game" example, where a group of people sit in a circle and tell each other a story following the circle, one by one each person either forgets, changes or embellishes a bit of the text and the last person then compares the original story to the last version and everybody laughs at how different the stories are.

This is not how oral tradition actually works or has worked for millennia, first, let me give you an example with yourself; remember a song, any song you know well, especially if you know it since childhood or since you were a teenager. You remember the melody, the key and the lyrics so perfectly that you can sing it unchanged twenty, thirty and even 50 years later.

So, the way oral tradition works is that a tribe or group chose someone or more than one person from the community to be their History book, this is usually a child and the previous historian begins training this child, the child is taught a chant, and then the next day the child will chant to the group, if the child makes a mistake, the child will be corrected and will be taught all of the chant again. The next day the child will chant again, as many times as necessary, until no mistakes are made; this process is incremental, after the child knows many chants, the child will chant to the group always beginning from the first chant incrementally. This training takes many years, and the chants will be recorded in the memory of this person until it is time to teach a successor. The worst tragedy was that this person died, because the history of a group/clan/family might be lost forever, this is why the master and apprentice overlap, and if the master dies the apprentice will replace him, also several members of the community/tribe know at least one chant, and if the apprentice dies, the master has time to train another before he dies.

This is why you must never dismiss oral history like (for example) the one of the Native Americans, this is serious business for them and the fact that it is not written doesn't make it less accurate or real.

There is another issue which should at least be mentioned, we tend to "feel" something written is different than a "told story", but please remember that people sometimes write things that are not true, intentionally or not. The fact that something is written doesn't make it true, just keep this in mind.

Returning to where we were, we now know there was one or many great floods in the past, cultures all over the world know or knew about it and passed the information down in their oral and written traditions; we have geological records of when some of these floods happened, and more geological evidence is discovered every day. Why would I say that there were many, simply because there were many global floods, but also many local floods, we will begin about 12.000 years ago, (Don't freak out creationists, we already agreed we had to mention the timeline of "the science guys" so they don't get lost), what we don't know is if one or all of them were regarded as global. The problem is that if you live on an atoll or island and sea level rises "your whole world" will be covered by water, you simply might not know there is a rest of the world that didn't flood.

To give you an idea, here is a graph of the latest great floods on earth and their magnitude. If you are a creationist, just replace 14,700 with 6,000 and be done with it.

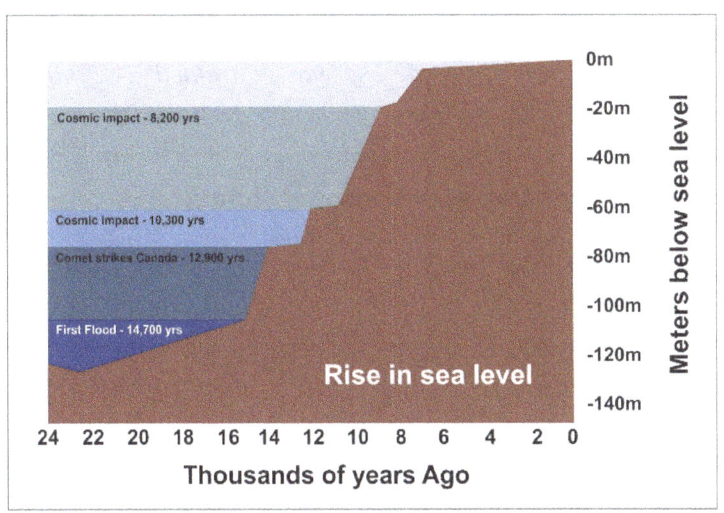

140 meters (or 500 feet) is a lot of depth, and these floods were not localized; these floods were overall sea level rise, so if you lived 10,300 years ago and sea level rose 40 meters, the flooding would have been catastrophic, 40 meters is about half a block of vertical water level, depending on where you live this might flood thousands of square miles of plains. There are several areas of the planet where the continental shelf is shallow, in these areas, massive areas would be flooded, as an example we can take Doggerland (also called Dogger Littoral) which was an area of land, now submerged beneath the southern North Sea, which connected Great Britain, Europe and Scandinavia, rising sea levels covered it about 6500–6200 BC. Not our flood, but some areas of Doggerland were above water for a long time after that time, until a mega tsunami caused by a landslide in what today is Norway hit the area and leveled it.

Now, I want to quote the following passage of the bible:

"The springs of the deep and the floodgates of the heavens were closed, and the rain from the sky was restrained. The waters receded steadily from the earth, and after 150 days the waters had gone down. On the seventeenth day of the seventh month, the ark came to rest on the mountains of Ararat..."

This passage is similar to: *"Then he stood very straight and tall, reaching as high as a cloud, and the flood waters came as high as he did. When he fell, the waters receded and there was dry ground..."* This is the flood myth of the Arnhem Land (Northern Territory of Australia)

Also we can read: *"There he gathered together wood and made a raft. Men and women and animals with him placed themselves. (Please note: Themselves) The mountaintop disappeared from their view, and they floated along on the waters. For many days they floated. At last, the flood began to subside. Soon the people on the raft saw the trees on the tops of the mountains. Then they saw the mountains and hills, then the plains and the valleys."* (Ojibwe – Native American)

Another Native American flood story says: *"A great inundation, together with an earthquake, swept the land so rapidly that only a few people escaped in their skin canoes to the tops of the highest mountains."* (Innuit – Native American)

Quaternary geologists, paleoclimatologists, and a whole bunch of other "...ogists" call the first flood in the graph Meltwater Pulse 1A, and it did happen about 14,700 years ago, this sea level rise event was believed to have taken 400 years to complete, however this does not match the stories of some of our ancestors or the geological record. Several geologists argue that the flood was much faster than originally thought.

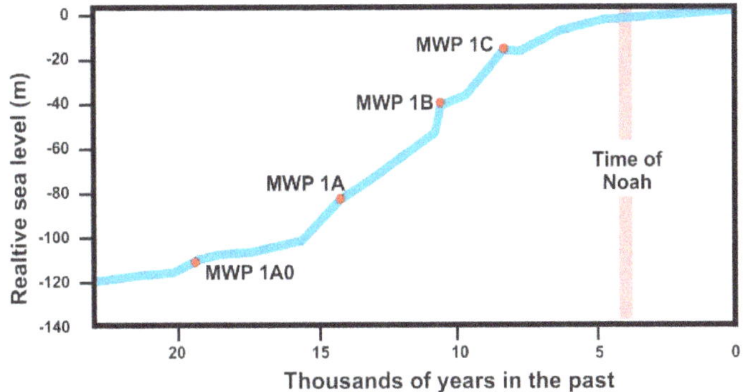

So, does Meltwater pulse 1A match a "great flood" myth in the past? It does, it matches several flood myths, like the one from the Tamil in India:

"Half of the land mass Kumari Kandam, which was south of India, sank in a great flood, destroying the first Tamil Sangam. The people moved to the other half and established the second Tamil Sangam there, but the rest of Kumari too sank beneath the sea. The lone survivor was a Tamil prince named Thirumaaran, who managed to rescue some Tamil literary classics and swim with them to present-day Tamil Nadu."

But this myth does not tell us about a great flood that after some time recedes, it is a two stage flood, sea level remains the same for a while and then another flood comes and covers the rest of the land with the water level remaining high, this correlates with the geological record for metlwater pulses 1A and 1B.

A great worldwide flood at the end of the last ice age that happened about 12,000 years ago. Now, that was a great flood, but it certainly does not seem to be Noah's flood. The end of the Ice Age was too long ago, not just because the Bible says so, but also because the great floods after the last ice age raised sea level by 400 feet, but the water level never came down again, the sea level stayed as it is now. That sea level rise flooded what we now know as continental shelves. Anthropologists, scriptures and myths tell us that humans (before and after the flood) were already developed Homo sapiens, (not brutes) so they couldn't have messed up the dates so bad, and wouldn't tell us that the water rose and then came down when it didn't. This prompted me to begin trying to find another worldwide flooding that must have happened about 4,000 to 4,500 years ago. So you see, I began with an idea, and I was wrong. I could act like a scientist (which I am not) and tell everybody "what do religious and native people know; sure they made a mistake in the dates". But I can't believe that people would miss the timeline by 7.000 years, and tell us that the sea level rose and dropped when it clearly didn't. Regarding the dates, it's simply too long ago and also it's the wrong flood (and very arrogant of scientists to disregard these stories in that way).

Atlantis

Please don´t think I'm insane because I mention Atlantis, I am mentioning it just as an example of "the wrong flood". If Atlantis existed, it is now underwater; therefore it can´t be the deluge described in the Bible, in the deluge the water level came down after flooding the land. Plato told the story of Atlantis, and this has been called a myth for centuries, but please remember the story of how Troy was found, people searched for Troy for thousands of years, and since it hadn´t been found, people began to think it was a myth, however, Heinrich Schliemann found it, how? By simply following Homer to the letter, he just read Homer and followed the landmarks. Now Plato is another story, first because Plato told the world a second hand story, and also because it was way before Plato, thousands of years before.

Plato tells the story of Solon, and how Egyptian priests, described to him Atlantis as an island larger than Asia Minor and Libya combined, just beyond the Pillars of Hercules (the Strait of Gibraltar). That in itself is odd, but when Solon asked the Egyptians when this happened they told him 9,000 years ago, and Solon went to Egypt, on 600 B.C. so that would add up to close to 12,000 years ago.

Image by David Foxx from Pixabay

The Egyptians told Solon that 9,000 years before, survivors of a land of seafarers that sank under the sea came to Egypt, bringing knowledge, civilization and teachings.

Image by Gianni Crestani from Pixabay

However, to pinpoint what happened, when and how, we must also understand the environment of North Africa. The Sahara desert undergoes dry and humid periods in a cycle, there is ample evidence of this cycle, the last humid period began about 14,600–14,500 years ago at the end of the Heinrich event and at the same time of the Bølling-Allerød warming, and according to calculations, the Sahara will cease to be a desert in about 10,000 years.

Photo: Depositphotos.com

So, during the last ice age, the Sahara was a desert, just like today, and humans lived in the Nile valley, which was quite densely populated. However in a short period of time, rain patterns shifted and the Sahara became green, researchers say that the change took less than 500 years, this allowed humans to migrate into the recently created savannahs, there was animal life, plants, lakes, rivers and it became a great place to live. It was at about this time when the survivors from Atlantis are said to have arrived. Please keep in mind that Atlantis has not been proven to have existed and that even its name is not certain, that is the name given to it by the Greeks. Two major dry fluctuations occurred in North Africa at about the time of the Younger Dryas.

The Younger Dryas, is a period of climatic disturbances on earth, named after a plant (Dryas octopetala) that grows in the cold, and that was common in Europe at the time. This period was a global cooling that returned the planet to Ice Age temperatures and conditions after considerable warming, Some researchers believe this was caused by dust hurled into the atmosphere by comets impacting earth, others believe volcanic ash from massive eruptions caused this, and other scientists believe that the massive influx of melt water from ice changed the salinity ratio of the oceans and current patterns causing massive temperature changes, the temperature record is there and there was a massive drop in temperature at the time.

Image by Thomas B. from Pixabay

The African humid period ended about 6,000–5,000 years ago during a cold period called the Piora Oscillation. So, once again, we find that major climatic changes did happen thousands of years before the time of Noah, but we don't see these major climate changes when we look for them at later times, like 4,000 years ago when the deluge is said to have happened. We have records myths and traditions telling about floods, but no major climate shift to explain them, temperatures on Earth remained fairly stable for the last 10,000 years, or for the last 6,000 years (for creationists), this makes a climate induced flood impossible.

So we have two or more floods, one that didn't recede and one or some that did, one 12.000 years ago and another 4,500 to 5,000 years ago, which according to science "didn't happen" it has been so hard for science to identify what happened due to "lack of evidence", however there is a reason for this, one is uniformitarianism.

If you type uniformitarianism on Google, this is what you will get:

"Along with Charles Lyell, James Hutton developed the concept of uniformitarianism ...

Uniformitarianism is the idea that Earth has always changed in uniform ways and that the present is the key to the past. The principle of uniformitarianism is essential to understanding Earth's history."

https://www.nationalgeographic.org/encyclopedia/uniformitarianism/

So based on uniformitarianism, catastrophism simply can´t exist.

If you type catastrophism on Google, this is what you will get:

"Catastrophism, doctrine that explains the differences in fossil forms encountered in successive stratigraphic levels as being the product of repeated cataclysmic occurrences and repeated new

creations. This doctrine generally is associated with the great French naturalist Baron Georges Cuvier (1769–1832)".

Britannica, The Editors of Encyclopaedia. "Catastrophism". Encyclopedia Britannica, 24 Jun. 2013, https://www.britannica.com/science/catastrophism-geology. Accessed 12 August 2021.

And uniformitarianism or gradualism dominated the scene for decades, so no global floods, super volcanoes, coronal mass ejections, comets or asteroid impacts. This meant that everything had to be slow, gradual and constant, the last ice age needed thousands of years to begin and end, the dinosaurs died off a few at a time, and Mammoths died out one at a time because they were hunted to extinction by humans. I guess ancient humans also hunted down more than 35 species of megafauna including giant sloths, sabre-tooth cats, mastodons and mammoths.

So, just like the extinction of mammoths "by humans", scientists also decided that everything religious was allegory or metaphors and that of course meant the flood didn't happen.

As a second reason, there was no interdisciplinary communication. Specialization is a great thing, people become very good at what they do or at what they research, but no one is looking at the big picture. Theologists new about this, anthropologists new as well, and of course geologists and paleoclimatologists were on the trail, but archeologists didn't communicate with any of them. An example of uniformitarianism at work can be seen in the life and work of J. Harlan Bretz and his work in Washington State in the Channeled scablands, a man who the geological community ostracized for decades, but that in the end, when Mr. Bretz was in his 90s, the scientific community had to accept he was right, he was awarded the Pemrose medal. A funny comment he made was that all those who ostracized him had died of old age, so he didn´t get to gloat over them.

An example of "interference" of one field of research into another was when Dr. Robert Schoch was invited to Egypt by Mr. John Anthony West to study the Sphinx.

Dr. Robert Schoch has taught at Boston University since 1984. An associate professor of Natural Sciences at the College of General Studies, He teaches undergraduate science courses, including biology, geology, environmental science, geography, science and public policy.

He received the Peyton Richter Award for interdisciplinary teaching. He is also co-author of the college textbook: Environmental Science: Systems and Solutions, now in its fifth edition.

In 1993, a genus of extinct mammals, Robertschochia, of which Robertschochia sullivani is the genoholotype, was named after him.

Dr. Schoch re-dated the sphinx in Giza, Egypt; to 10,000–5,000 BC, based on water erosion marks he identified on the Sphinx enclosure walls, and also on findings from seismic studies around the base of the Sphinx and elsewhere on the plateau.

This was not welcome by archaeologists and Egyptologists who attacked him viciously; the important part for us is that Schoch found water erosion in the sphinx enclosure, water erosion caused by massive flooding between 10,000 and 5,000 BC. But Dr. Schoch was also caught in this timeline, because he says there were no floods or rainfall in Egypt after 5,000 B.C. But if we allow ourselves to regard Noah's flood as real, we will see that it is possible to have had massive floods in Egypt 4,000 to 4,500 years ago or about 2,000 to 2,500 B.C. if Noah's flood actually happened when the Bible says it did. If this is the case, this doesn't sound so strange, does it?

In Schoch's own words, *"this was not rainfall during millennia it was massive water flows and rainfall in a short period of time"*.

So, if the sphinx was built when Dr. Schoch says it was, that would place its construction before the flood, making it much older than the pyramids, and we can see that the pyramids don´t show water damage similar to that shown by the sphinx enclosure. Although they might have been built and later repaired. In fact, we don´t see water erosion damage (or damage similar to the one the sphinx enclosure shows) in any of the Egyptian monuments or buildings. If this is the case, then we can attribute the damage to a short massive flood event, since the environment and climatic conditions of Egypt were not able to generate rainfall at the time.

I do have to mention that Dr. Schoch has his own explanation for some of these phenomena. He advocates coronal mass ejections or solar outbursts causing climate disturbances, maybe the end of the last ice age and massive civilization collapse.

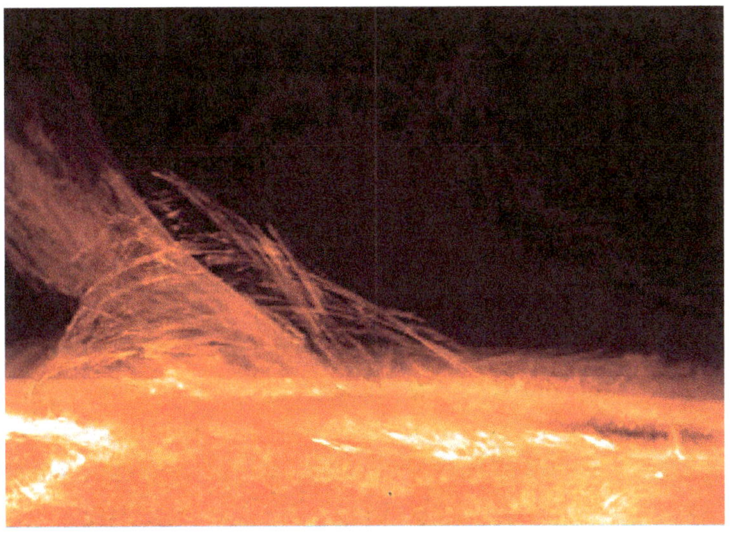

Image by WikiImages from Pixabay

Dr. Schoch gathers evidence from several scientific disciplines; he explains how the last ice age ended abruptly in 9700 BC due to solar coronal mass ejections. These solar outbursts caused electrical/plasma discharges to hit Earth and triggered volcanoes, seismic movements, wildfires, and massive floods as glaciers melted and lightning strikes released torrential rain by evaporation of the oceans. He tells us how these events eradicated the civilization of the time and set humanity back thousands of years, only to reemerge around 3500 BC with scattered memories. He explains how many megalithic monuments, underground cities, and ancient legends fall into place, as well as the reinterpreted Easter Island rongorongo texts and the intentional burial, 10,000 years ago, of the Göbekli Tepe complex in Turkey. *(From the description of his book Forgotten civilization).*

Dr. Schoch dates the sphinx at the time because he genuinely believes there was no flood 4,000 years ago, but is this a fact? According to Dr. Schoch, Mr. John Anthony West, Mr. Graham Hancock, Mr. Randall Carlson and others, there was a civilization before 9.700 BCE, How is this relevant to Noah and his flood?

Well, very simple; Noah knew about the flood before it happened, and it is possible that the survivors from that long gone civilization passed down knowledge on how to predict these global events. You see, if you believe God warned Noah, it's ok, but if you are reading these lines and you are an atheist, then the survivors of that civilization could have been the ones to pass down the knowledge that warned Noah and others like him.

Now, returning to creationism and non-creationism, there is an explanation as to why the world would begin 6,000 years ago, and if you are not a creationist and think that a comet impact or a solar outburst caused the end of the last ice age and the fast melting of the ice, causing mass extinctions and global destruction; you must agree that humanity had to begin once again "as children", so you must allow for a few thousand years for humanity to recover from the massive impact of loss of fertile soil (which is now underwater). And if you are a creationist, after all the destruction and death, the world actually began 6,000 years ago.

I read all of the flood myths I could find, some seem to talk about meltwater pulses 1A and 1B, others talk about flooded lakes, but many talk about a great flood, and receding waters, so I started to search for a flood that could have affected the Middle East roughly at the time of Noah's flood. It seems I was looking at the wrong flood when I began this.

The Sumerian flood myth seems to fit Noah's flood, it reads:

"The gods had decided to destroy mankind. The god Enlil warned the priest-king Ziusudra ("Long of Life") of the coming flood by speaking to a wall while Ziusudra listened at the side. He was instructed to build a great ship and carry beasts and birds upon it. Violent winds came, and a flood of rain covered the earth for seven days and nights. Then Ziusudra opened a window in the large boat, allowing sunlight to enter, and he prostrated himself before the sun-god Utu. After landing, he sacrificed a sheep and an ox and bowed before Anu and Enlil. For protecting the animals and the seed of mankind, he was granted eternal life and taken to the country of Dilmun, where the sun rises."

And the Assyrian myth fits even better:

"The waters of the abyss rose up, and it stormed for six days. Even the gods were frightened by the flood's fury. Upon seeing all the people killed, the gods repented and wept. The waters covered everything but the top of the mountain Nisur, where the boat landed. Seven days later, Utnapishtim released a dove, but it returned finding nowhere else to land. He next returned a sparrow, which also returned, and then a raven, which did not return. Thus he knew the waters had receded enough for the people to emerge."

This is an extract from a Native American legend:

"...Nanabozho, fleeing before the angry waters, thought of his Indian children. He ran through their villages, shouting, "Run to the mountaintops! The Great Serpent is angry and is flooding the earth! Run! Run!"

The people caught up their children and found safety on the mountains. Nanabozho continued his flight along the base of the western hills and then up a high mountain beyond Lake Superior, far to the north. There he found many men and animals that had escaped from the flood that was already covering the valleys and plains and even the highest hills. Still the waters continued to rise. Soon all the mountains were under the flood, except the high one on which stood Nanabozho.

There he gathered together timber and made a raft. Upon it the men and women and animals with him placed themselves. Almost immediately the mountaintop disappeared from their view, and they floated along on the face of the waters. For many days they floated. At long last, the flood began to subside. Soon the people on the raft saw the trees on the tops of the mountains. Then they saw the mountains and hills, then the plains and the valleys.

When the water disappeared from the land, the people who survived learned that the Great Serpent was dead and that his companions had returned to the bottom of the lake of spirits. There they remain to this day. For fear of Nanabozho, they have never dared to come forth again."

And this is from the Ottawa:

The flood raged on; but, though mountains of water were continually being hurled after the prophet, he was safe. When he had floated on the water many days, he ordered Aw-milk (the beaver) to dive down and, if he could reach the bottom, to bring up some earth. Down the latter plunged, but in a few minutes came floating to the surface lifeless. The prophet pulled him into the boat, blew into his mouth, and he became alive again. He then said to Waw-jashk (the musk-rat), "You are the best diver among all the animal creation. Go down to the bottom and bring me up some earth, out of which I will create a new world; for we cannot much longer live on the face of the deep."

Down plunged the musk-rat; but, like the beaver, he, too, soon came to the surface lifeless, and was drawn into the boat, whereupon the prophet blew into his mouth, and he became alive again. In his paw, however, was found a small quantity of earth, which the prophet rolled into a small ball, and tied to the neck of Ka-ke-gi (the raven), saying, "Go thou, and fly to and fro over the surface of the deep, that dry land may appear." The raven did so; the waters rolled away; the world resumed its former shape; and, in course of time, the maiden and prophet were united and repeopled the world.

There are hundreds of these stories all over the world, and many of these refer to a serpent, a snake, a fire spewing dragon, stars falling, angels falling, and fire from the sky.

The Pilaga (pit´laxá), who are part of the Guaicurúes of the Great Chaco, and are closely related with the Toba, both inhabit the center of the province of Formosa in Argentina in South America today, have a myth about a day when parts of the moon came crashing down on Earth, according to them this was thousands of years ago, and tell us there was a war between the sun and the moon, and that the war made coal, hot enough to melt sand to fall from the sky. The fire from the impact is said to have burned the whole world (at least the world of the Pilaga), and that people were burned alive where they stood, but that some tribes became alligators and hid inside the lakes. These impacts actually happened and they are called "Campo del Cielo", and have been dated to 4,000 years ago, or 2,800 B.C.

The Qom myth considered the "rocks" (asteroids) to be drops of the sun's sweat; Thus they were worshipped by saying that when dawn broke on clear days they were transformed into erect tree trunks (there seems to be here, in the structure of this myth, the notion of a fertilization of the Earth from solar events), likewise the Qom and neighboring ethnic groups such as the Mokoit and Abipones assumed the Campo del Cielo area as a place where the Earth and the Sun met intimately.

The Wichí, who live today in Argentina and Bolivia tell a myth that is different; maybe because the Wichi were not direct witnesses of the meteor shower and learned the myth from their neighbors, transforming it according to their belief system. For the Wichis, the meteoric shower had occurred when the jaguars attacked the moon, taking away some pieces of it.

Although less perceptible, also in this myth there seems to be a reference to the fertilization of the Earth, in this case from the effects of the moon. Another explanation for the difference between the Qom and Wichi myths may have been due to the fact that the ancestors of the Qom, were closer to the site of the meteor shower, perceived the asteroids as large objects of a color and brightness similar to those of the sun, while the Wichi would have seen a large shower of "shooting stars" of a "silvery" color similar to that of the moon.

A Chinese myth tells us about a sea monster named Gonggong, it seems that 5,000 years ago there was a battle between Gonggong and the God of fire named Zhurong for the throne of heaven (or the sky); when Gonggong lost the war, in a fit of rage he hurled himself into Buzhou mountain which the people of the time believe it was one of the 8 pillars that held up the sky. The impact broke the sky and caused storms of flames and devastating floods. Most stories say that Gonggong caused the heavens to tilt toward the northwest, and the sun, moon and stars move in that direction.

So, a scientifically proven asteroid impact at about 4,000 years ago in South America, and an impact 5,000 years ago in China that caused global flooding with massive tsunamis.

As modern human beings we all know where this is going.

But how can we find the right ones?

Well, today I was on my way home, I went to buy some groceries and there had been a traffic accident, not too serious, but the police was there, a patrol car and a police van. This van was a traffic accident investigation unit; I stopped and asked one of the officers about the accident, these people are sort of like Crime Scene Investigators for traffic accidents, he told me exactly what happened, he collected witness accounts, studied the tire markings, the damage to the cars, and developed an exact picture of how the accident happened. I then decided to borrow his method, as I hopped back on my motorcycle to go home, I felt I had a way to find out exactly what happened thousands of years ago.

So let's use the same method to investigate this, we just arrived to the accident, it is true the accident was thousands of years ago, but the accident left markings, we can see the results, we have the witness accounts in the form of myths and scriptures, we have massive chevrons in different areas of the world, which also can be dated, marine skeletons inland, layers of deposits that can be analyzed to determine their composition, so we will begin by cataloguing all the witness accounts to determine which ones are from our accident and which ones are from other accidents that will confuse our findings, we are essentially interviewing witnesses of the accident.

Well, let's think about this, Lucifer is a fallen angel something like and in the garden of Eden a serpent was the personification of evil, the devil, and fire...

The Mesoamericans had Quetzalcoatl/Kukulkan, as I already mentioned Quetzal has a long feathered tail and Coatl is a serpent, the Chinese used to equate comets with dragons or to "long tailed pheasants", so...

The investigation

First I plotted a map of the Earth with all the flood myths I could find and read, this map shows these flood myths in their actual location, wherever the myth came from I added a dot and I marked the dots in two colors, I assigned red to the ones that say that the water came and then receded, and blue to the myths that say that the water rose and never came down. Of course we are now assuming all these myths as true accounts of what happened, and regard them as witness accounts. Please see it the way I do, why would the people from Papua, New Guinea or the Australian aborigines lie about the history of their people to themselves?

If you look at the map you will find a distribution of these floods.

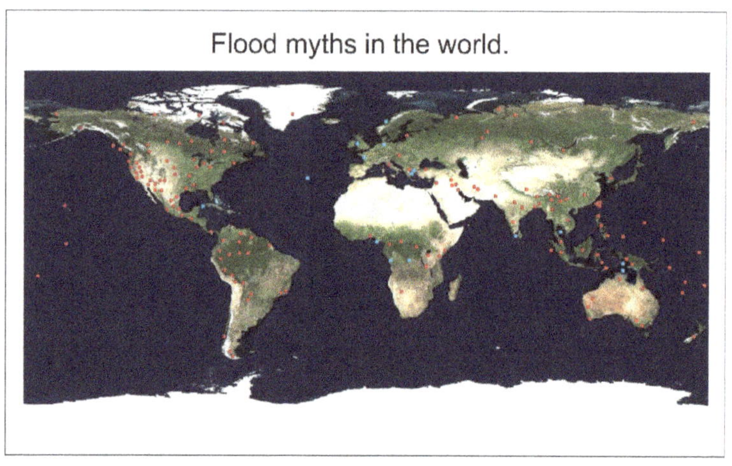

Public domain map of the Earth (with my dots added)

Most of these floods are red, very few of them are blue.

For scientists, this flood thing has gone from "It never happened", to "there were some floods", to "This must have been the flooding from the end of the last ice age, but still it was not a global flood". Now, if you look at the red dots, these are the floods where the water rose and came down, we can't say that these are all the same flood as the blue dots, the blue dots are the myths and stories which tell of a flood where water never receded, like the one from the Tamil and the one of Atlantis, then if we add that to the global distribution of the flood myths and that we are told by scripture and myth that there was a global flood, this seems to fit as a preliminary hypothesis.

Also, since we know that meltwater pulses 1a and 1b happened around the Younger Dryas, we can (as per this hypothesis) just ignore the blue dots as belonging to a previous flood or floods. The premise being that the floods where the water level never came down were earlier than the one we are looking for and were caused by a totally different event.

Some of our witnesses are well known, such as Plato and his story of Atlantis, but I will not cover that story, when Solon went to Egypt (Kemit) it was 600 B.C. and the Egyptian priests of the temple of Edfu told Solon that the story of Atlantis (yes it was Egyptian, not Greek), happened 9,000 years before, we add 9,000 + 600 + 2021 and we get 11,621, and the end of the last ice age with its massive sea level rises and cataclysms happened at... Yes, you guessed it, about 12,000 years ago, this flood is not the one we are looking for; we are looking for one that must have happened at about 4.300 years ago. Why this timing for the flood we are looking for?

Well, because as I mentioned before, Christian creationists date the creation of the world at about 6.000 years ago, and the great flood is calculated to have occurred about 1.650 years after creation, so if we are looking for Noah's flood we need a flood that fits, at least roughly, the time mentioned in the bible.

This is the way Troy was found, by reading Homer, so if we want to find Noah's flood we must use the Bible to find that particular flood. After all we are looking for the biblical flood, not just find any flood and make it fit; it HAS to be Noah's flood.

Several of the ancient flood myths make references to waves or waters rising catastrophically, also some myths mention water surging from the ground, the earth shaking, the sun disappearing, darkness, rain, etc. The Bible mentions several things that match the ancient flood myths from other cultures.

Going on a bit of a tangent here, Imagine the surprise of the Spanish missionaries in the 1,500s when they met the people of the Hopi nation and they found out that their religion said...

"Tawa destroyed the Third World in a great flood. Before the destruction, Spider Grandmother sealed the more righteous people into hollow reeds which were used as boats. On arrival on a small piece of dry land, the people saw nothing around them but more water, even after planting a large bamboo shoot, climbing to the top, and looking about. Spider Woman then told the people to make boats out of more reeds, and using island "stepping-stones" along the way, the people sailed east until they arrived on the mountainous coasts of the Fourth World."

In Oraibi (the oldest of the Hopi villages), little children are often told the story of the sipapu, and the story of an ocean voyage is related to them when they are older. Even the name of the Hopi Water Clan (Patkinyamu) literally means "a dwelling-on-water" or "houseboat".

According to Barry Pritzker, *"In this Fourth World, the people learned many lessons about the proper way to live. They learned to worship Masauwu, who ensured that the dead return safely to the Underworld and who gave them the four sacred tablets that, in symbolic form, outlined their wanderings and their proper behavior in the Fourth World. Masauwu also told the people to watch for the Pahána, the Lost White Brother".*

So, the missionaries told the Native Americans their flood stories and the Native Americans said "This is true", and told them theirs. Of course, the missionaries said the flood destroyed the world shortly after its creation and the Hopi answered back surprised. Don't you know about the two worlds before that one?

Of course the missionaries didn't, but the Hopi understood that, because several neighboring tribes did know about the flood, but not about the previous worlds.

How would God go about sterilizing all life out of a planet, totally flooding it?

According to what we are told in the bible, it rained for 40 days and 40 nights, and then the planet was flooded for 150 days, that is the simple version, however the water cycle tells us that to have rain you need to evaporate water and that wouldn't cause the mountains to get covered with water, this would have required additional water, because rain does not raise sea level, the water comes from the sea when it evaporates.

Weather events, like storms, tornados, hurricanes are never global, our planet has a complex atmosphere and it will not support such events at a global level.

Also, all these witnesses had one cause in common, man was wicked and God decided to punish them (by drowning them all, and all animals as well), if God did such a thing, and God is an all-powerful being who expresses anger, satisfaction, or other emotions using weather, whatever God used to express his displeasure with humanity must have come from the sky (where God's things come from).

Now we have identified a set of floods, but we still need to find out which ones among all the ones we identified were part of a global flood; of course there might be some local ones that may have gotten mixed up with the global one, for example an earthquake caused a tsunami in certain island 6 years before our flood, then we couldn't tell them apart, but still, we can't say that all the floods shown in the map are local and happened at about the same time, that wouldn't make sense either.

I always remember a story about calculations, one of my teachers began doing math in his head, he wasn't being precise, he would mutter: about 4.6 times about 7.8 by 56.7 and then wrote down the result, we redid everything with a calculator, using all the decimals and the result was correct; when I asked why it was correct if he used inaccurate figures to calculate it, he answered: Because errors tend to cancel each other, you never make all your mistakes in one direction, sometimes you estimate too large, and sometimes too small, therefore the result, although not accurate, will not be that far off.

There is another thing we need to identify, this is the mechanism by which this happened, and we already talked about the impact event that killed off the dinosaurs, so you might have caught where this is going.

Fire From the Sky, Sound Blasts and Earthquakes

So I went back to the handy list of flood myths to check if the witnesses saw something significant in this regard. If I could find myths that said it rained for a month, or a moon, or 40 days and nights I would have been very happy, but that is not what I found, so I reread Genesis and found:

"…on that day all the springs of the great deep burst forth, and the floodgates of the heavens were opened. And rain fell on the earth forty days and forty nights."

So, it wasn't just rain in Noah's flood, the *"springs of the great deep burst forth"*

This might take the shape of fire from the sky, stars falling, angels falling massive clouds and sound, thunder or something sounding like it, and blasts or massive shockwaves and the like.

From India we get Brahma saying: *"…preserve all creatures and plants aboard a ship which had been prepared. Said that a hundred years of drought and famine would begin this day, which would be followed by fires from the sun, and from underground that would consume the earth and the ether, destroying this world, the gods, and the planets. Seven clouds from the steam of the fire will inundate the earth, and the three worlds will be reduced to one ocean.*

From North Central India: *"…He sent a stream of fire-water (Sengle-Daa) from heaven, and all people died save a brother and sister who had hidden beneath a tiril tree (this is why tiril wood is black and charred today)."*

From China we get a 22 year flood and: *"...Before Gun was finished however, the Supreme Sovereign sent the fire god Zhu Rong to execute him for his theft."*

From Santal (Bengal) we get: *"...Fire fell and devastated the country, destroying half the population."*

"...While people were at Khojkaman, their misdeeds became so great that the creator Thakur Jiu sent a fire-rain to punish them. Only two people escaped, in a cave on Mount Haradata."

"...The first people became incestuous and unheedful of God or their betters. Sirma Thakoor, or Sing Bonga, the creator, destroyed them, some say by water and others say by fire. He spared sixteen people."

"Some say by water, others by fire". Could it have been both?

The fire in or from the sky appears in more than 90 flood myths, also the sound or noise appears in many of them, like in the one from India: *"...They told Kabitt and Aka that in five days, when the moon was full, the sea will make a booming sound, and they should escape to a mountain where there are stars."*

"They told Kabitt and Aka that in five days, when the moon was full, the sea will make a booming sound". So they knew, just like Noah knew what was to come, although Noah new years in advance.

The wind or hurricanes were there too, we get from North Borneo: *"…In the middle of the night, the drum began sounding "Duk Duk Kagu" on its own. Then a great wind came and swept away all the houses, with the people in them. Some were carried out to sea;"*

And from Lithuania: *"…He sent two giants, Wandu and Wejas (water and wind), to destroy earth."*

From Cook Islands: *"…A wind burst forth, and the sea swept over the island of Rakaanga. A few inhabitants survived by taking refuge on a mound."*

From Tahiti: *"…Even the trees and stones were carried away by the wind. But two people were saved."*

From the Tchiglit Eskimo (Point Barrow to Cape Bathurst): *"…A great flood broke over the land. Driven by the wind, it submerged people's dwellings."*

Massive hurricane type winds appear in more than 40 of the flood myths, we could go on for days reading about the fire from the sky, the wind, the sounds, rain, and of course the earthquakes, more than 13 of the myths have either earthquakes, cracks on the ground where fire or water come out, like the Quechua flood myth from South America: *"…The stars also knew from the moon's spotted face that they were descended from an incestuous relationship. They all cried, and their crying produced rain, earthquake, and flood."*

From the Cheyenne in North America we read: *"…One very hard winter had earthquakes, volcanoes, and floods which destroyed all the trees. The people spent the long winter in caves and were almost famished the following spring. The Great Medicine, in pity, gave them corn and buffalo. Since then, there have been no more famines or floods."*

And from Peru: *"…Long ago, before there were any Incas, the country was populous, but the ocean broke out of its bounds, the land was covered, and the people perished. Some say that a few people survived in the caves of the highest mountains. Others say that only six people survived on a float."*

So, not only these "Myths" help us get a picture of what actually happened, but also that this was not just rain; furthermore, in several of the myths, people end up living in caves, food is scarce and people have to fight beasts for food.

We have been told by mainstream archaeology that "cavemen" were something like sub-human, which they were not, of course I am not an archaeologist or an anthropologist and I might be wrong; First of all, here we go back, to before the creationist date of creation, the scientific timeline. I am basing myself on what mainstream scientists say, according to them, we (as humans) have had a "modern size" brain since 500,000 years ago, and our DNA has been "modern" since 260,000 to 350,000 years ago.

To make a long story short, I believe "cavemen" were not actually "cavemen", they were people with a culture, knowledge but were recovering from a cataclysm.

To explain how this could have happened, there is a joke about a man that says: "I wish I could go to the past and give them all my knowledge", the person travels to the middle ages and tells everybody of electricity, phones, radios, etc. Then the people form the past ask "How do we make this electricity?", and the man answers "I don´t know".

The ancient ones and the warning

A common theme in most myths and in the bible is the messenger or God itself warning of what is to come.

Noah was warned by God that the flood was coming, also remember that *in India "the incarnation of Lord Vishnu as a fish forewarns Man…"*

In Taiwan "An old white-haired man came to Oppehnaboon in a dream and told him that a great storm would soon come." What a coincidence, I read in the book of Enoch, that Noah was born with white hair, you should really read the book of Enoch, I will not quote it much here, but it is interesting.

And the warnings go on, in Greek mythology: *"Deucalion, advised by Prometheus, built a chest. Stocked it with provisions, embarked in it with Pyrrha. Zeus, poured heavy rain from heaven, flooded a large part of Greece to destroy all men."*

So, we see that in most of these myths those who survived received a warning from God or messengers from God. I am going to attribute this warning to the massive trauma humans suffered from the floods of the end of the last ice age; Mainstream archaeologists have told us that we come from cavemen, that cavemen were cavemen because they didn´t know any better, they were nomadic, stupid hunter gatherers and did not have a language or a culture. However, although we are taught this, Homo Sapiens is described by the same scientists as rising about 300,000 years ago followed by anatomically modern Homo sapiens at least 200,000 years ago, and our brain became what it is today about 100,000 years ago. As an example, the cave paintings in Lascaux, France are dated at 17,000 – 15,000 BC. So, If we listen to science, these people were as smart as us, therefore not ignorant brutes without brains! I didn't say it, archaeology, geochronology and paleontology researchers said it, so they should explain this, since what is called "The great leap forward", a sudden emergence of technology,

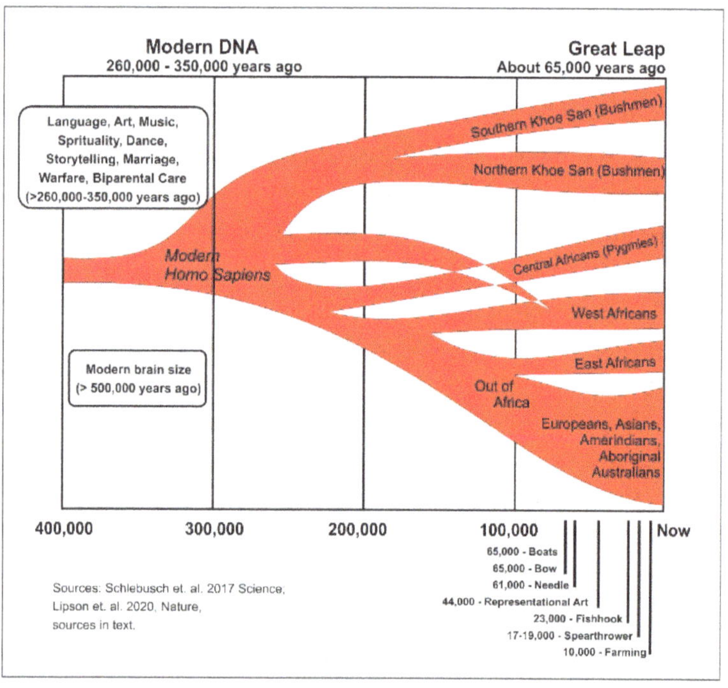

reflecting the evolution of a fully modern human brain happened 65,000 years ago. Also, they tell us that human intelligence could be described as modern far earlier.

So, we have the same scientists telling us that cavemen were brutes, hunters and gatherers, with no culture and they also tell us that "Modern Humans" had their "Great leap forward" about 65,000 years ago. If we remember the date of the end of the last ice age, it was 12,000 years ago, so it was 40,000 years after the "Great leap forward". According to these same scientists, language appeared between 350,000 and 260,000 years ago. So, they have some explaining to do if they want us to believe that humans sat around grunting at each other (or speaking to each other about nothing relevant) and doing nothing other than hunting, killing off every single mammoth, gathering and living in caves for 500,000 or 600,000 years and all of a sudden, 5,000 years ago decided to build cities, begin farming, and civilize themselves, maybe they got bored of having nothing to do.

I believe that people at the time of the Ice age were not uncivilized, but that they had a fair amount of knowledge about the world, its climate and astronomy, and they managed to record this for posterity.

We know that people lived in caves at the end of the last Ice Age, I believe they lived in caves because their homes and their land were destroyed and they needed to live in caves to get shelter and protection from the catastrophe raging outside.

"The remains of hundreds of humans from diverse ethnic groups, who died from natural causes in caves (apparently seeking

refuge from a catastrophe), have been discovered in North America, Brazil, India and the Balkans."

http://gurdjiefffourthway.org/pdf/ATLANTIS%20AND%20EGYPT.pdf

I will not cover the possibility of civilizations which seem to have existed at the time of the Ice age or before, but if the people at the time (and this one is the flood that is said to have happened 14,000 years ago at the end of the last ice age) suffered an extinction level cataclysm, and it was a worldwide flood in which the water level never came down, and they saw what caused it, and according to these same scientists, they already had language, art, music, spirituality before 200,000 to 500,000 years ago, I am sure they would have told their descendants about it.

Ancient astronomers (astrologists at the time), recorded stellar events very accurately and these have been confirmed with modern observations, for example, the earliest known supernova, named HB9, was recorded in India in 4,500 BC (+/- 1,000 years), Comets, asteroid impacts, meteor showers, and other events have been catalogued and recorded.

Let's keep in mind that these people had an in depth knowledge of our skies, it has been said that these people had nothing else to do but look at the sky, that they had clear unpolluted air and that the sky they saw was way more spectacular than the one we see today; however, you have to keep in mind that although they were very proficient in different activities they did burn wood for heat, they had unpaved roads, and also had volcanoes and other pollution, so extremely clear skies might not be the actual explanation.

They didn´t see the world like we do, many of these people had either living Gods or had Gods that actively interfered in their lives, moreover; the night sky was not just something over them. To take the Greeks as an example, their Gods were jealous, cheated on their wives (which were Goddesses) the Goddesses cheated on their husbands, and mated with mortals. The night sky was like a telenovela for them, an ongoing drama of their Gods. As an example, just think of the story of Orion, who fell in love with Pleione and his daughters and followed them to see if he could win their affection.

The Pleiades, in Greek mythology, the seven daughters of the Titan Atlas and the Oceanid Pleione: Maia, Electra, Taygete, Celaeno, Alcyone, Sterope, and Merope. They all had children by gods (except Merope, who married Sisyphus).

Atlas and Pleione had seven daughters, Maia, Electra, Taygete, Celaeno, Alcyone, Sterope, all but Merope who married Sisyphushad and had children with Gods, The Pleiades then formed or were made a constellation by Zeus. According to a myth they all killed themselves out of grief when their sisters, the Hyades died. Another myth states that after seven years of being pursued by a Boeotian giant named Orion, they were turned into stars by Zeus. Orion also became a constellation, and continued to chase the sisters across the sky. Merope, the faintest star of the Pleiades was thought to be ashamed of falling in love with a mortal, but it could also be Electra, grieving for Troy, the city of Dardanus, who was her son with Zeus.

So you see, jealousy, intrigue, love, cheating, and actual persons who were fathered by the Gods who could be identified in real life down here on Earth. Any meteorite, any movement in the sky, any minute change in the arrangement of the stars or constellations was tracked and recorded, because their Gods were an integral part of their lives and the sky told them what the Gods were up to.

A comet would bring famine, death or destruction; so it was tracked when it appeared and left, everything was so important that had to be watched, if they worked out the period of a comet, they would record it and take precautions for the next time it came, this could be storing grain or sacrificing a young man or a virgin; this we will never know, but the Gods must be placated or else. Just imagine if Poseidon got angry, then a tsunami would come, so people watched the night sky and if they saw something that might anger Poseidon, they would record it and warn everybody that the God of the sea might act up.

I am just trying to get you to understand their mindset, try to stand in their shoes, a comet came and went, then it came and went again, this over many generations, then they would know that it would come right after or right before the winter solstice every (let's say) 157 years. They were not interested in the comet itself, but maybe in that it crossed the bull's face and annoyed it, and maybe the next time it passed the bull might charge it and hurl it into the Earth.

Maybe a "small" body, like the Chelyabinsk meteor that entered Earth's atmosphere over the southern Ural region in Russia on 2013 seemed to come from Orion and these people would think that Orion was angry and shot an arrow down to Earth.

Or, imagine this, they knew that Jupiter passed the sky on a certain path, they also knew that there is a periodic comet (flying serpent) passing that year and they will cross paths, depending on if Jupiter was in a good mood, maybe because his cousin Mars was close by, it could grab the snake and stop it from passing over or hitting Earth, on the other hand, if Jupiter was in a bad mood it could grab the snake and hurl it down to Earth and kill off millions.

In 2003, Mars was closer to Earth than in the last 60,000 years, this will not happen again until 2,287. When Mars and Earth are close, Mars appears very bright and for the Romans this meant the god of war as close by, this could either warn them not to get into war or might get them to start one. Ancient people 60,000 years ago saw a big orange star moving across the sky, what they thought it would bring us will never know, but they knew about it, saw it and recorded it, and be sure they knew exactly when it was going to happen again.

This is not limited to celestial bodies, this also applies to weather. Mainstream science has led us to believe that these people did not know or understand what a storm was, and they made up Zeus (or any other past or future thunder and lightning God) to explain them; and that might be true to a certain extent, but once that happens, then this God has a temper, He has struggles and things that upset Him. So, look at it this way, the God sends lightning and thunder every time He gets upset; it is true that these people didn't know what a thunderstorm is or what caused it, but they knew what happens immediately before a thunderstorm, maybe the wind God blows towards the high mountain, or the wind comes from the sea, and every time the wind God does this, the thunder God becomes angry. We today look at a synoptic chart and see a low pressure area and know rain is coming, they saw their Gods doing things and that helped them predict what the other Gods would do. Apply this to tornadoes or draughts or whatever weather event you want.

Yes, to them a lightning was something the sky God threw down to Earth, they didn't know electricity and how it was caused, but they knew that if the wind God was upset and blew across the land in the summer and left until next year, then his brother the lightning God would be upset because He loved him very much and will not see him until next year, and hurl lightning down into their village.

I just made up this mythology, but you get the point. It was a different knowledge, it helped them with weather forecast, and the night sky with the prediction of other events, such as comets, meteor showers, Planetary wobbles, earthquakes, etc. Since they were their Gods, and they saw them as "hands on" Gods, the ancient people kept very precise records of what their Gods did in the past, in case they did it again.

Keep in mind that these people were well aware and kept track of our planet's precession cycle. There are two precessions for our planet: Axial and Apsidal, Apsidal precession lasts about 112,000 years and changes the orientation of our orbit relative to the elliptical plane. The combined axial and apsidal precession result in a precession cycle of about 23,000 years. They didn't know why, nor the physics involved, but that changed the solstices, changed climate, rain patterns, and they knew all about it, because their forefathers recorded it.

The Australian aborigines will tell anyone who will listen about how their people walked from the north into Australia. This has already been verified by genetic studies, Australian aborigines arrived to Australia walking from the Sunda (today Borneo, Sumatra and Java) and then, into new guinea, then walking to Australia between 65.000 and 50.000 years ago, please keep in mind this is the oldest uninterrupted culture on the planet. Because after the end of the last ice age, the land they walked through to get to Australia was flooded and it has remained underwater since.

Some of these stories from Australia have been dated, and belong to a flood in which water never subsided.

Port Phillip Bay

Numerous tribes described a time when the bay was mostly dry land, when the bay "was a kangaroo ground.", "Plenty catch kangaroo and plenty catch opossum there." Researchers determined that these stories are from a time when sea level was about 30 feet lower than today, about 7,800 to 9,350 years ago.

Kangaroo Island

The Ngarrindjeri people tell stories of Ngurunderi, a mythological character. In one of their stories, *Ngurunderi chased his wives until they fled to Kangaroo Island on foot. Ngurunderi angrily rose the seas, turning the women into rocks that now are above surface between the island and the mainland.* This tale originated at a time when seas were about

100 feet lower than today, dating the story at 9,800 to 10,650 years ago.

Tiwi Islands

A story told by the Tiwi people describes the mythological creation of Bathurst and Melville islands in the North. *An old woman is said to have crawled between the islands, followed by a flow of water. The story is interpreted as the settling of what now are islands, followed by flooding around them, which seems to have happened 8,200 to 9,650 years ago.*

Rottnest, Carnac and Garden Islands

An early European settler heard Aboriginal stories telling how these islands, in the coast of Perth or Fremantle, *"once were part of the mainland, and that the ground there was forest." According to at least one story, the trees caught fire, and that the ground split under the forest with a great noise, and the sea rushed in between, cutting off these islands from the mainland."* Based on the bathymetry, the story dates back 7,500 to 8,900 years ago.

Fitzroy Island

Stories tell of a time when the shoreline stretched so far out that it abutted the Great Barrier Reef. The stories tell of a river that entered the sea at Fitzroy Island. The great gulf between today's shoreline and the reef tells of a time when sea was 200 feet lower than today, placing the story at 12,600 years ago.

Spencer Gulf

Spencer Gulf, now underwater, was a floodplain lined with freshwater lagoons, the stories told by the Narrangga people. This could have been between 9,550 and 12,450 years ago.

But more people moved to Australia at a later stage. Prof. Mark Stoneking, from the Max Planck Institute for Evolutionary Anthropology - Department of Evolutionary Genetics points out that: *"We have a pretty clear signal from looking at a large number of genetic markers from all across the genome that there was contact between India and Australia somewhere around 4,000 to 5,000 years ago."*

This time period matches the fall of several Middle Eastern civilizations, could this migration have been caused by the destruction of the mega-tsunami of Noah's flood? Great events cause migrations, and several cultures were seafaring ones. Now, why move from India to anywhere else if nothing happened? When land is flooded by salt water the soil is destroyed, you can't grow anything on it for a long time, and 4,000 to 5,000 years ago was about the time of Noah. The dating is not precise because these are studies based on changes in DNA, but it gives us a hint that something very important happened at the time, not only because of the migration, but because it only happened once (that we know of).

However, millennia later the evidence of the migration was found by modern technology.

The study analyzed DNA from 344 people; it was published in the journal Proceedings of the National Academy of Sciences, and DNA samples were taken from other populations to see if Indian DNA was found there, but since no Indian DNA was found in other populations in Asia, this has led scientists to believe Indians sailed directly on boats through the Indian ocean. Wild dogs seem to have arrived at the same time.

Dr Irina Pugach, of the Max Planck Institute for Evolutionary Anthropology in Leipzig, Germany, believes the Indian DNA reached Australia 141 generations ago, assuming each on average lasted 30 years the geneticists were able to calculate that the Indian population arrived to Australia 4230 years ago.

So, why did they migrate, was it just because they were explorers?

If that was the case, they wouldn't have stayed; they would have stayed for a short while and left, however they did stay, and there were many of them, enough to leave a genetic fingerprint in the local population. This, and the date of the migration, leads me to believe that they left India because the aftermath of the deluge was so large that they had to migrate and find another place to settle, of course they couldn't know that the event was global, but they must have found a place in Australia which wasn't affected so badly, maybe protected by the coastal highlands, settling in the central area; the Northern territory or South Australia. Current estimates of the aborigine population in what is modern day Australia is about 3.000.000 people and since modern aborigine Australians are estimated to have about 11% Indian DNA, those who came 4,000 years ago were not just a few individuals, the migration could not have been small to alter the DNA of 3 million inhabitants (people living in Australia at the time).

We can assume these people migrated from India because of the effects of the Burckle impact which must have polluted the soil and the tsunami that came later salted it leaving it barren.

I keep moving around the world, because the flood stories are spread all over it, and they match; however, in Europe, cave paintings are a clue of how knowledgeable these people were about their surroundings, for example, the cave paintings of Lascaux in France seem to be a depiction of at least part of the night sky, and show Orion, Taurus and even the Pleiades.

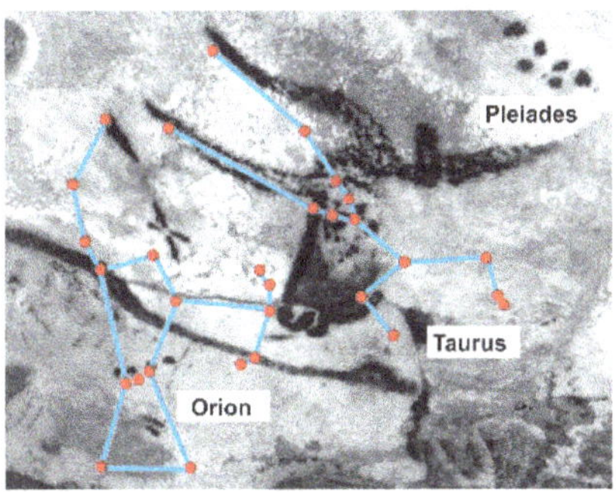

When Pablo Picasso was invited to see the cave paintings in Lascaux he is quoted saying "We have invented nothing".

Massive Megalithic sites were constructed before and after the last Ice Age as astronomical observatories, and although archaeologists say these had "agricultural" significance, as in: "Let's build this to mark the solstices so we know when to sow our crops"; I believe they are wrong, or at least only partially right. They told us that people have been as smart as us for the last 300,000 or 65,000 years, (maybe they should be a little more precise), so why would a society build something like Stonehenge or monuments like it to figure out the solstices when literally 2 big rocks or 2 sticks will do the job? These people were not dumb.

This would be like these ancient people building animal corrals with multi-ton rocks, it just doesn't make sense, these people were obsessed with the night sky, they required very precise readings and there must have been a reason for it.

Also, although sites like Stonehenge have been dated to 3,000 BC, there are other sites like the 9,000-year-old underground megalithic settlement of Atlit Yam, or the several megaliths that are underwater, as well as whole cities, like one in the gulf of Cambay (India), the 9,500-year-old city, called the Dwarka Kingdom which is 5 kilometers long is said to have been the city of the Hindu God Krishna, and the site called Gobekle Tepe in Turkey, but I digress.

Jumping to the other side of the world, specifically to Central America, the Mayan people predicted the return of Quetzalcoatl/Kukulkan for between 1993 and 2012, Quetzal (long tailed bird) and coatl (Rattlesnake) seem to represent a comet, the Chinese called comets "Long tailed pheasant stars", and the Maya called the Pleiades "Tzab-ek" (the rattles of the rattlesnake), also the quetzal and the rattlesnake are depicted in green in the Mayan codices; In 2004 a green comet flew past the Pleiades and its tail passed in front of the Pleiades, which gave the rattlesnake its rattles. So, as you can see, these "Ignorant tribesmen" weren't actually as ignorant as scientists might think. Floods and storms are associated with Quetzalcoatl/Kukulkan. On the other side of the world the Romans and Greeks had the messenger of the gods Mercury/Hermes, (a green comet) also associated to floods and storms.

The point is that the constellations of Taurus and Orion seemed to have been very important for these ancient peoples.

So now we have this to keep in mind The Mayan people predicted the return of Quetzalcoatl/Kukulkan for between 1993 and 2012", and they were right.

The Taurid meteor stream

If I told you to blindfold yourself and walk right through a busy highway you would laugh, but that is precisely what the Earth does twice a year. *"This stream produces at least four meteor showers on Earth: the Northern and Southern Taurids, both active from end of September until December; the Daytime ζ-Perseids, active from end of May to the beginning of July; and the Daytime β-Taurids, active in June and the first half of July (Jenniskens 2006). Other showers may be also related to the Taurid stream, namely the Piscids in September, χ-Orionids in December, and Daytime May Arietids in May (Jenniskens 2006)"*

https://www.aanda.org/articles/aa/full_html/2017/09/aa30787-17/aa30787-17.html

Comet Encke/P is supposed to be the origin of this meteor stream, but it has also been proposed that this comet used to be part of a larger body which fragmented thousands of years ago (guess what, astronomers estimate, 10.000 years).

If the end of the last ice age was caused by a comet or asteroid impact, these ancient people passed down the message or warning and we can see it in Egypt, according to Graham Hancock and Robert Bauval:

"...we have demonstrated with a substantial body of evidence that the pattern of stars that is "frozen" on the ground at Giza in the form of the three pyramids and the Sphinx represents the disposition of the constellations of Orion and Leo as they looked at the moment of sunrise on the spring equinox during the astronomical "Age of Leo" (i.e., the epoch in which the Sun was "housed" by Leo on the spring equinox.) Like all precessional ages this was a 2,160-year period. It is generally calculated to have fallen between the Gregorian calendar dates of 10,970 and 8810 BC." **The Mars Mystery - 1988**

Back in present times, on February 15, 2013, an airburst (when a meteorite hits the atmosphere but does not reach the ground) over the city of Chelyabinsk, Russia, likely caused by a stony asteroid or meteorite the size of a five-floor building that broke apart at an altitude of 24 km sent a shockwave equivalent to a 550 kiloton explosion, blew out more than a million windows and injured more than a thousand people in six Russian cities.

550 kilotons is 36 times the atomic bomb Little Boy dropped in Hiroshima, and this was a small asteroid which did not get even close to hitting the ground.

A couple of years ago, a new comet was discovered headed towards Earth, the object was originally named 2014 UN271. It's now been officially named Comet Bernardinelli-Bernstein as its discoverers; the current estimate suggests its nucleus, or core, is between 62 to 230 miles (100 to 370 km) in diameter. Now, if we remember that the comet or asteroid that killed off the dinosaurs was 10 kilometers (6.2 miles) in diameter, then this new one is actually a planet killer. Don´t worry, it will not come even close to earth, it seems it will circle the sun at about the distance of the orbit of Saturn. So, we are safe, but keep in mind that it is at a minimum, six times larger than the one that killed off the dinosaurs, but it could be 23 times larger. An impact like this would simply sterilize Earth, maybe blow off its atmosphere or cause it to explode. Yes THERE ARE big things out there.

As a reminder, at first I thought that Noah's flood was simply the abrupt end to the last ice age and that the description of the flood was inaccurate, but after reading many, not all the flood myths; But a lot of them, I had to change my assumption, Noah's flood sounds more like a mega tsunami, where the water came and left, the large amount of rain also points to an enormous atmospheric disturbance concurrent with the flood, according to The Bible, water also came out of the ground, so we can see earthquakes and upheaval of the water table as part of the picture.

The only release of energy, if we discard an underwater explosive supervolcano. Which is large enough to be capable of causing a global flood or global tsunami, is an impact of a very large body (or bodies) on the planet. So far, geologists have not found evidence of supervolcanoes erupting about 4,000 years ago. The most recent eruption of a supervolcano on Earth known to geologists occurred 27,000 years ago at Taupo located at the center of New Zealand's north island, so this does not seem to be the cause of Noah's flood. If a supervolcano was the culprit, we would have found the evidence of two of them, one from 27,000 years ago and one from 4,500 years ago right on top of each other.

We have confirmed only one mass extinction event due to an impact, and this is the impact that killed off the dinosaurs about 65 million years ago.

Just as a reminder… There is geological evidence of an asteroid of about 10 km (6 miles) hitting Earth about 65 million years ago (once again, science timeline). This impact caused a huge explosion and carved a 180 km (roughly 110 mile) crater on what is today the shore of the Yucatan peninsula. Material from the explosion was pulverized and thrown into the atmosphere, severely altering climate, and causing the extinction of about 75% of all species from that time, including the dinosaurs. We know of many asteroids like this one; their orbits enter the inner solar system and intersect Earth's orbit. Some could hit Earth in the future. Most, but not all of them are smaller than the one that hit Earth 65 million years ago.

You don't have to be a genius to figure out that periodical meteor showers hit Earth or that comets appear periodically; you just have to look up to the sky, keep good records, and some large event like the one that ended the last ice age (whatever it was), would be included in all oral records and warned about. If this event was a comet impact, it could have broken down in parts, some of them hitting different parts of the planet and some continuing their orbit; our ancestors would have recorded and calculated when it was going to happen again, maybe they saw the impacts and saw a large part of the comet that just flew by and left, and the Maya wisely predicted the return of Quetzalcoatl.

The 1994 impact on Jupiter

Please let me tell you about something that happened a few years ago, and for the first time in human history we could watch, measure and observe, not just the scars of an impact, but the impact itself and the effects that are still ongoing. Comet Shoemaker–Levy 9 was a comet that broke up in July 1992 and impacted Jupiter in July 1994, and this was the first time we could watch the collision of a comet and a planet live.

Schematic of broken up Shoemaker–Levy 9

The volume of all the fragments of the comet (fifteen) would add up to a 1.8 km diameter comet. Assuming it's density was in the range of 0.5 g/cm^3, Shoemaker-Levy 9 had a probable total kinetic energy in all of the impacts equivalent to 300 Gigatons of TNT. According to research, the US and Russian nuclear arsenals combined have a yield of about 6,600 megatons, this means you would need 45 times the yield of the total nuclear arsenal of the world to replicate the force of the impact on Jupiter. So, no; global nuclear war doesn't sound as bad as a comet impact.

Whenever we imagine an impact event on the planet, we imagine one big rock and a massive crater, but the Shoemaker–Levy 9 impact on Jupiter showed us that if a comet hits earth things might be quite different. As I already mentioned this comet broke up before hitting Jupiter, so what actually hit Jupiter was a volley of 15 small comets, so if this would happen on Earth, given that the earth spins at 460 meters per second, or about 1,000 miles per hour, imagine that these fragments arrived to Earth (for the sake of argument) at about 1 every half an hour, this would mean fifteen impacts separated by 500 miles each, depending on the angle and the direction, more fragments could hit land or sea, but the event would be devastating and it is doubtful humanity would even survive it.

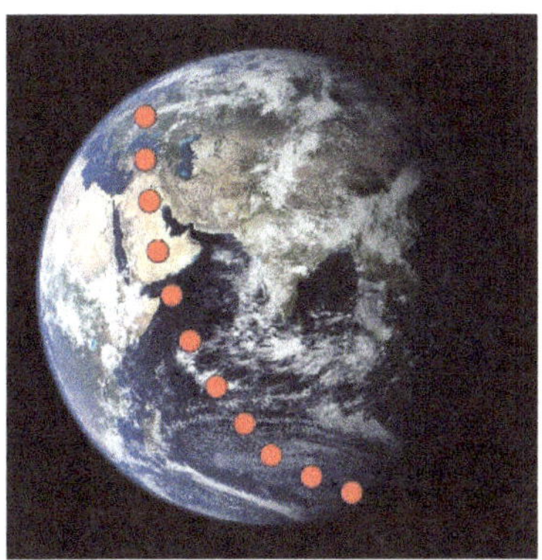

For as long as we can remember, there have been significant impacts on the Earth's environment.

The 1908 Tunguska object, about 50 meters in size, hit over Siberia, producing a fireball as bright as the Sun and eventually exploding above the ground in the Tunguska River region. The energy of the impact was close to 20 megatons, a thousand times the energy of the Hiroshima bomb, and flattened more than 2,000 square kilometers of forest, 18,000,000 trees were flattened by an iron object that exploded over Russia's Sikhote-Alin region in 1947 was the second largest known impact of this century, creating more than 100 craters 1 to 14 meters in diameter.

Astronomers have calculated the average frequency with which impacts of different sizes can occur. For example, an impact the size of Tunguska is expected every 100 years, while an object large enough to produce the crater detected in the Yucatan Peninsula (Mexico), and believed to have killed all the dinosaurs, occurs maybe every 100 million years.

To simplify this, astronomers often assume that these impacts are completely random. We can ask ourselves if this is true or if there are time periods when there are many impacts and others when there are fewer. These variations during human history are interesting.

Anyone who watches meteors knows that they do not arrive at random in time. Some nights a year the Earth passes through a stream of meteors and we call this a meteor shower, more meteors than average on some nights than others. Occasionally, the Earth passes through an unusually dense concentration of particles and a meteor storm occurs. The meteors that we see are small particles that burn up in the atmosphere.

There are reasons to believe that larger objects (the size of Tunguska and above) are close together in space, they seem to travel in packs or in lines, so that the Earth sometimes travels through a group of these objects. This is based on the idea of large comets disintegrating or fragmenting due to gravity forces.

The collision of the fragments of comet Shoemaker-Levy 9 with Jupiter in 1994, not only showed the world that comets can impact planets, but also that fragmenting of comets can happen and produce volleys in a short period of time.

Medium-sized comets are more than the really big ones; and a giant comet contains more mass than all its fragments. Therefore, it is the largest comets (over 100 km in diameter) that are the most significant in terms of disturbing the Earth's environment. In general, one such giant comet is captured in the inner solar system every 100 thousand years (but this can vary). This giant comet will then split and fragment over many thousands of years, but its parts will travel together, maybe in a line, one after the other, or orbiting each other.

There seems to be a giant comet whose parts have been present throughout human history, and are still present in near-Earth space. This suggests that the Taurid Meteor Stream is the remnants of an extremely large comet.

If the original body, before fragmenting was closely related to comet Encke (just as an example), it appears that the last intersections were about 300 to 500 AD, and before that before 2000 BC. Although there is uncertainty in these calculations. The most recent intersections could have been a couple of centuries earlier, and the earlier ones could have been as early as 3000 BC. Although we don´t have precise dating resulting from these calculations, these always show that there are "dangerous periods" lasting a couple of centuries, in which there is a large increase in impacts, followed by a few thousand years in which the Earth's orbit is not intersected by the Taurid Meteor Stream.

The fear of comets of many cultures in history may be based on occasional fragmentations or Tunguska type impacts. Several cultures have legends of fireballs falling from the sky that are associated with fires and deaths. Gildas, a British priest recorded a catastrophe in Britain in the 5th century AD, after which there was a large migration of people to northern France. This time is close to the decline and fall of the Roman Empire and the beginning of the European Middle Ages, when temperatures plummeted, there was an excess of rain and crops failed in Europe.

Dr. Duncan Steel, an astronomer; suggests that many ancient monuments, including Stonehenge in Britain and the pyramids in Egypt, might be related to the existence of an unusually active sky at that time, the pyramids are believed to be closely related to Orion, which happens to be beside Taurus in the night sky. He has shown how the past orbit of the Taurid Meteor Stream of the time could also be related to the original orientation of Stonehenge.

Image by Sally Wilson from Pixabay

Let's assume that the shamans, priests or wise men, knowing that these meteor showers came every year, and that periodic comets did exist, they went and talked to a resourceful citizen like Noah and told him to be vigilant, but if we keep in mind that the Maya knew the cycle of the green comet, and the Romans knew it too, I think they narrowed it down to a decade or a few years, even calculated the orbit and knew when the impact was going to happen, then went to the resourceful guy to tell him to start building. Also, these shamans or wise men, who were avid observers of the night sky would notice differences (Like a new star or snake in the sky), actually a comet which lights up in the night sky, and if they knew it was going to hit because it seemed to be coming head on or was going to be a very close fly-by, they would go warn people about this; they didn't even have to calculate the orbit, in some cases, depending on the knowledge of the culture, they could have warned months or years in advance just based on observations.

A lot has been said about aliens warning humans about these things "the gods", but let's keep in mind that human intelligence has been just like ours for some time. The event called the "Great Leap Forward" which led to modern behavior rapidly increasing sophistication in tool-making and behavior seems to have happened between 80,000 years ago and 60,000 years ago. Totally modern behavior, such as figurative art, music, self-ornamentation, trade, burial rites etc. appears (or we have found evidence of it) about 30,000 years ago. Now think that we have gone from an agricultural society to putting a man on the moon, and now close to putting a man on Mars in less than 300 years. With this, I don't want to say that there was a civilization like ours before, although there was certainly time for it to develop, but keep in mind that ancient Greeks calculated the distance from the earth to the moon, and were pretty close.

The earliest records of calculating the distance from Earth to the Moon were by Aristarchus of Samos in the 4th century BC and later by Hipparchus, these calculations gave a result of (376,000–427,000 km or 233,000–265,000 mi).

Modern figures, at perigee (closest), the moon comes as close as 225,623 miles (363,104 kilometers). At apogee (farthest), the moon is 252,088 miles (405,696 km) away from Earth, averaging 238,855 miles (384,400 km).

To calculate these things you need very precise observations, you need to know exactly where you are standing, the direction in which you are looking and where the bodies you are observing are, this explains something that many archaeologists and Egyptologists have no explanation for; why align a pyramid so precisely to true north if it is just a glorified tombstone? Pyramids all over the world are precisely aligned and located; the Mayan pyramids have observatories on top of them, you can see, still today the sun align perfectly with the top of a tower in Angkor-Wat.

Using computer models, an amateur historian and an astronomer (Kinsman and Asher) have found evidence that many important events recorded in Mayan hieroglyphs match with meteor showers caused by remnants of Halley's Comet.

In newly published research, the team has found more than a dozen hieroglyphs from the Mayan Classic Period (250–909 CE) showing important events within just a few days of Eta Aquariid meteor showers, related to the comet. It seems that the Mayans could predict meteor showers related to comets. Who is to say that if they were aware of the periodicity of Halley's comet, that they were not aware of other similar ones with periods of thousands of years.

Yield of an impact

The impact that killed off the dinosaurs is widely accepted today, but the proponents of it were not that well regarded in the past, no matter how much geological evidence was collected, the "theory" was ridiculed, attacked and debunking attempts were constant. The science is settled now, but a massive extinction of that size and so fast needs to hit hard and have global consequences. We will now use that impact as a model to calculate how much damage a comet or asteroid could cause.

The University of Texas calculated that the impact generated close to a billion times the energy of the atomic bombs dropped on Hiroshima and was a million times the largest nuclear bomb ever tested, "The initial crater was close to 100 kilometers wide and 30 kilometers deep."

The Imperial College said it blasted high velocity material into the atmosphere, causing a global winter, wiping out much of life on Earth in days."

The asteroid hit what is today the Yucatan peninsula, yes, Cancún, where the resorts are today. Paleontology documents the extinction of the dinosaurs as well as more than 50 percent of plant and insect species.

"Combining all available data from different science disciplines led us to conclude that a large asteroid impact 65 million years ago was the major cause of the mass extinctions," - Peter Schulte, assistant professor at the University of Erlangen in Germany.

The Earth Impact Effects Program website created by Gareth Collins shows the extent of the fireball that engulfed the impact zone and the area covered by the shockwave.

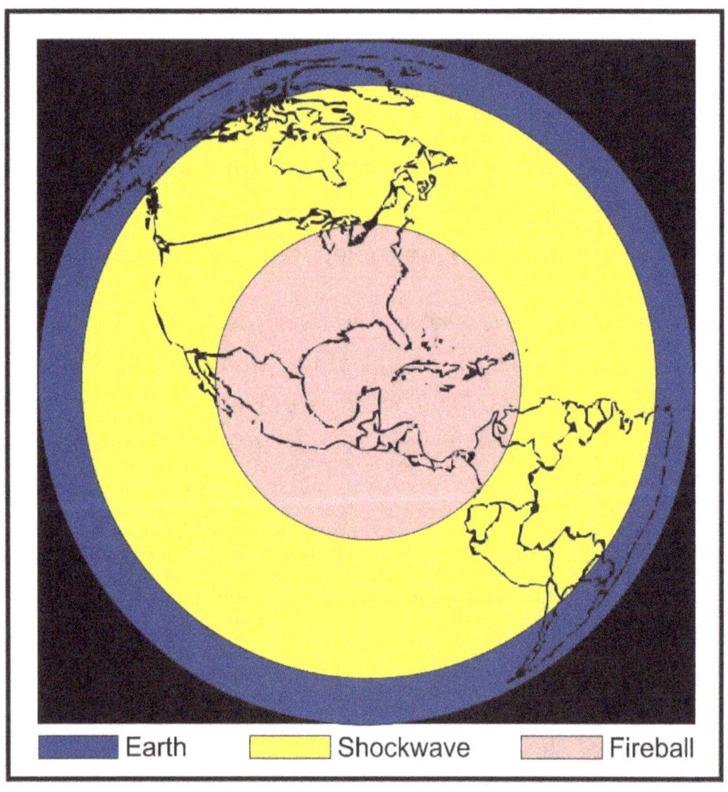

I am not going to bore you with a detailed description of the blast that killed off the dinosaurs, but some information is necessary, the pink fireball area you see above is about 5.000 kilometers (3.100 miles) in diameter, this is an area where literally the air caught fire. The yellow area affected by the shockwave or air blast is roughly 10.000 kilometers (6.200 miles) in diameter. The shockwave of the impact that killed off the Dinosaurs was global, and also the tsunami that followed.

"…one thing is very clear: The Chicxulub tsunami was clearly a force to be reckoned with." As Arbic said, *"It must have been one of the biggest tsunamis ever."*

—Katherine Kornei (hobbies4kk@gmail.com; @katherinekornei), Freelance Science Journalist

Citation: Kornei, K. (2018), Huge global tsunami followed dinosaur-killing asteroid impact, Eos, 99, https://doi.org/10.1029/2018EO112419. Published on 20 December 2018.

Another example for a reference yield would be a scenario of a 10 kilometer comet hitting Earth, it would hit at about 70 kilometers per second, which would be about 250,000 kilometers per hour, or 156,000 miles per hour, but it would hit the atmosphere first, a comet hitting the atmosphere would be like an atomic bomb exploding in mid air. After heating the air to temperatures high enough to vaporize rock, the comet would impact the ground, just about 1.5 seconds later, creating a crater that would be close to 100 kilometers (60 miles) in diameter and close to 30 kilometers (19 miles) deep. The air of shockwave would travel at unimaginable speed, while the impact would raise the temperature of the rock under the impact to 2,370 degrees C, or 4,298 F, roughly half the temperature of the surface of the Sun. Inside a radius of 1000 km (600 miles) from the impact point, clothes, people, trees, wood all ignite, then get flattened by a shockwave that uproots forests whole.

I just wanted to give you an example, but the effects of this kind of shockwave on a much smaller scale could be seen in Tunguska, Siberia. That impact had a yield between 15 and 30 megatons and the shockwave or blast area was 2.000 square kilometers, about the area of Mauritius.

What did it?

Let's move forward in time, millions of years, we are now at a different time, mammals now rule the earth, dinosaurs are long gone, and humans, the dominant species don't even know about dinosaurs yet, the ice age ended millennia ago and other than some myths and legends nobody remembers the last impact that ended the ice age, but in the darkness of space something is coming towards Earth, the size of this comet is nowhere near the size of the asteroid that killed off the dinosaurs, but it is still large and it's coming.

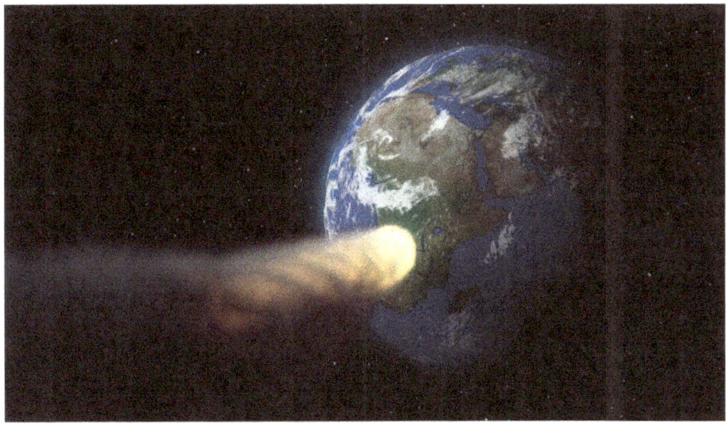

Image by MasterTux from Pixabay

While the general population of the time minded their own business; others, with knowledge of astronomy had already spotted the comet, shamans, priests, astrologists and those who knew of the ancient legends were concerned and began warning of the coming disaster. We will never know if people like Noah were warned by someone or knew themselves what was coming, we only know what was told to us in myth and scripture that Noah was warned directly by God, others by messengers, fish, animals, beings; take your pick, but the fact is they knew about it, some were told to get to high ground, others were told to hollow tress, others to tie trees and build rafts, Noah was told to build an ark. We can also see that the farther away from the source, more people survived, even complete tribes.

Yes, there was a source, the main suspect is a recently found impact crater at the bottom of the sea, between Madagascar and India, it is the Buckrle Abyssal Impact Crater, and it is a crater with a diameter of 29 kilometers (18.1 miles) located at 30.87° S 61.36°E, but for most of us coordinates mean nothing, so this actually means between India and Africa, to the south, closer to Africa than India, in front of Madagascar.

This is just the moment to show you the following myth: A native story from Chile tells us of "… *2 great serpents who wanted to see which of them could make the sea rise more, causing an earthquake and sending a great wave onto the shore.*"

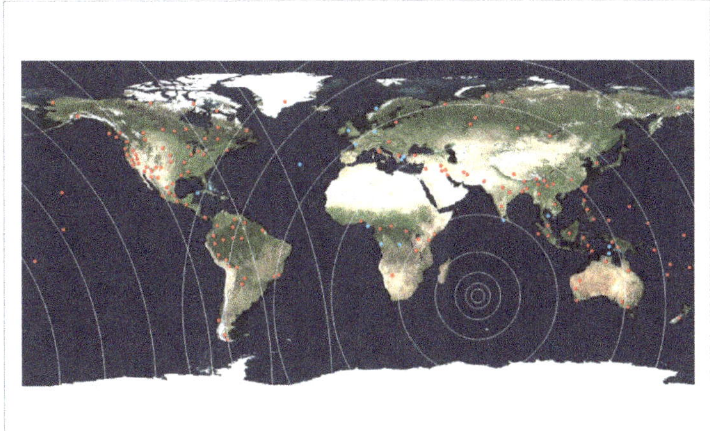

I have added these circles radiating from the crater to the map of Flood Myths. If we keep in mind that the fireball of the Asteroid that killed off the Dinosaurs was about 5.000 kilometers in diameter and the crater was close to 100 kilometers wide, and we are pretty sure the Bruckle impact was caused not by an asteroid but by a comet, because people were warned that it was coming, and Noah had enough time to build a big boat, (it must have been visible for a long time before impact).

In the Bible, Noah is warned by God, although in the book of Enoch he is warned by someone else, we will discuss that later, for now we will go with God. How long did god give Noah to build the ark?

Well, according to answergenesis.org there is an estimate of the time:

So Noah had between 55 and 75 years of warning before the deluge came, plenty of time to build his ark.

The most important thing to point out about our flood myths (which I have already called witness accounts), is not what we find, but what we don't find. We don't find any flood myths in Madagascar or South East Africa, this area must have been within the kill zone and no one was left alive to tell the story. The diameter of the Burckle crater is about 29 kilometers, therefore we can estimate a very large kill zone or fireball, let's assume 1,000 to 1,500 kilometers in diameter (about 1,000 miles), Burckle crater is 900 miles from Madagascar, which in turn is 267 miles from Lake Rukwa (just north of Lake Malawi), where we find our closest African flood myth, 1,425 miles from the Burckle crater.

Some will say that there was no one living there, but people did live in Africa at the time, you can see in the following map that the flood myths are recorded in Africa, but only if the people who recorded them lived far enough from the source of the event.

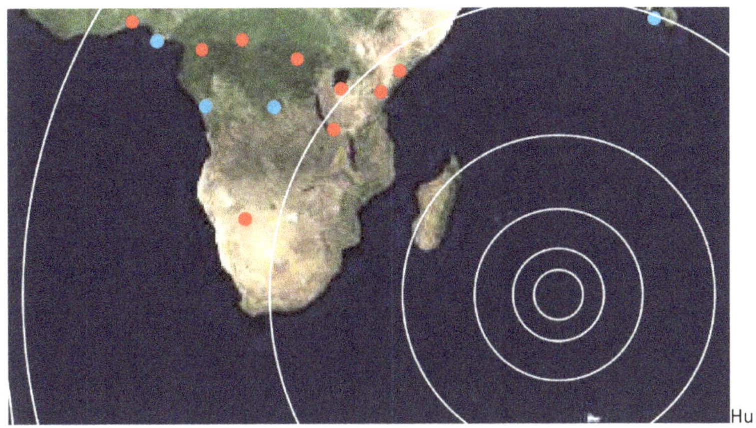

mans in Madagascar and the South East part of Africa were killed off by the Burckle impact, so there are no flood myths in the area close to the impact.

The burckle impact left a 29 kilometer (18 miles) scar in the bottom of the sea, east of Madagascar and west of Western Australia in the southern Indian Ocean at a depth of 3,800 meters (2.375 miles). What is also interesting is how it was found.

It is now that I have to mention the work of Michel Tellinger, a pharmacist who has very farfetched ideas about the southern part of Africa; I will not judge his work, but although some of his theories are extremely "out there", the places he has researched and catalogued are absolutely impressive. Stone circles, stone enclosed areas, and thousands of square miles of what he calls "agricultural terraces" in southern Africa. The area is called Mpumalanga, and features the Blaauboschkraal stone ruins, originally declared a national monument on 18 April 1975. These ruins are some a number of stone circles located in the Mpumalanga escarpment in an area of about 150 km^2.

"There are many disputes regarding the age of these structures, ranging from 25,000 to 250,000 years old. Needless to say, the Bakoni ruins have never been excavated to date—let alone thoroughly researched."

http://pseudoarchaeology.leadr.msu.edu/bakoniruins/

According to mainstream science these are the remains of structures created by the Bokoni people who settled the region in the 16th century, and they were built to be used mainly as cattle enclosures.

According to Tellinger, these structures are more than 200,000 years old, and could not have been cattle enclosures because they have no doors, gates or entrances or exits, but other than discuss their age and purpose, we see several thousands of them, with evidence of large scale farming in the area in the past.

You can see them on Google Earth from *"Ohrigstad to Carolina and connected over 10,000 square kilometers of the Mpumalanga escarpment into a complex web of walled structures"* (Schoeman).

"One of South Africa's most extensive and remarkable legacies of the past is little known by the public and largely ignored by heritage authorities" (Schoeman).

Anyway, these are ruins, of people, a civilization, culture or tribe who are not there anymore, many of them covered by the soil (so at least some of them might be thousands of years older than others), my point being that if this area was excavated or some boreholes drilled, they should be able to find strata with evidence of the Burckle impact. This area was hit hard by the shockwave, since it is in Africa, South of Madagascar, and was in the "Line of sight" of the impact. Research like this could help date the structures, if they are above this impact proxies layer, they would be less than 4,000 to 5,000 years old. Now if some of the ones which lie underground happen to be below that layer, or right on it; the impact could explain why these people don´t live there anymore. The area would have been devastated by the shockwave/fireball, and the dust settled on it would have ruined the soil, killed the plant life and of course incinerate anyone who might have been living at the time.

So far, nobody takes Michael Tellinger's explanations or theories seriously, but he has done something important, he has made these sites known to the public and this by itself might trigger research on them, which would in turn help mankind find out who lived there and when.

Finding the crater

The Holocene Impact Working Group (HIWG) is a group of scientists that are convinced that meteor and comet impacts are far more frequent than it was thought before, but before you run to buy or build a bomb shelter, it is still not that often, some scientists have gone from every several million years to hundreds of thousands to tens of thousands but still are in the "more than 1,000" years, this means that we don't have to walk around looking at the sky every time we go to get groceries, but this also means that at least some astronomers should be scanning some areas of the sky, because our ancestors did watch their sky, told us to, and they knew what they were talking about.

The group is composed of scientists who are experts in their respective fields

They self describe as an *"Ad hoc group called the Holocene Impact Working Group (HIWG) was created as follow-up the Workshop on Comets/Asteroid Hazard that was held in the Canary Islands in December of 2004. The group includes the researchers from different field of geo-science who believe that Holocene comet impacts were more frequent in the recent past than the accepted view. "*

As I already mentioned, the Holocene Impact Working Group (HIWG) was made aware of mega tsunami chevrons in different continents and these travelled long distances inland, so the group used satellite images of the chevrons to plot their direction, (where they seemed to come from and go to). They did this in Madagascar, Australia, Africa, and India, plotting direction lines from all of them, they saw that all these lines met in one point in the ocean, and that is where they looked.

Geologists Jody Bourgeois and R. Weiss challenged the origin of the chevrons in 2009, they stated that their models identified the chevrons as originated by wind; however, the fact that the Holocene Impact Working Group managed to triangulate the position of the Burckle crater using the directions of these chevrons, seems to confirm their origin. There is also another possibility, that the supersonic wind shockwave also played a part in forming the chevrons, so you see, sometimes both sides of an argument can be correct.

There is also a scientific research paper that relates this impact crater to flood myths, although I wasn't aware of the existence of this paper when I began writing these words, it is nice to find scientific corroboration to my assumptions.

Burckle Abyssal Impact Crater: Did this Impact Produce a Global Deluge? - *Dallas H. Abbott, Lloyd Burckle, and Perri Gerard-Little - Lamont Doherty Earth Observatory of Columbia University, Palisades, NY 10964 - W. Bruce Masse - Los Alamos National Laboratory, Los Alamos, NM 87545 - Dee Breger - Drexel University, Philadelphia, PA 19104.*

After separating the flood myths in two sets and hypothesizing that (although I wasn't even aware of Burckle crater yet), that there must have been an event at about the time of Noah and that the flood that was described in the Bible was not the same flood that was caused by the event that ended the last ice age.

So, let's go back to the impact, Noah is finished with his boat (Ark), he has been told to prepare, all over the world, several Noah's are doing the same thing, preparing for what is coming.

And it hit, with the force of hundreds of atomic bombs, you could pile all the nuclear arsenal of the planet in Burckle crater and detonate it and it would still be a minute fraction of the power of this explosion, it hit the atmosphere at a speed of more than 20 times the speed of sound, the initial blast incinerated everything in the surrounding area, nothing survived inside a 1,000 mile circle around the impact; but it didn't hit on land, it hit water. The impact site is heated up to 8,000°C, melting the bedrock of the bottom of the sea, and sending molten rocks flying at supersonic speed, The blast immediately vaporized billions of tons of water and seabed, Africa, the Arabian Peninsula, India, mainland Southeast Asia, Indonesia, and Australia, more than 9000 km (5.600 miles) away from the crater were blanketed by ejecta (debris from the impact) the supersonic shockwave hit the closest landmasses in minutes, People in India saw the plume from 5.000 kilometers (3,000 miles) away, described it as doomsday clouds *"...that look like a herd of elephants, emitting lightning, roaring loudly.... These dense, elephant clouds filled up the sky...."* Indian oral myths and texts (Yes, the Indians wrote about this event) describe fire particles falling from the sky at the beginning of the deluge, myths from the Congo region of Africa and New Guinea, describe the same *"Fire falling from the sky"*, other myths from Iraq, India, and New Guinea tell about hot water, actual hot rainfall falling from the sky.

Image by DeSa81 from Pixabay

Millions of tons of gravel and dust have been thrown into the upper atmosphere, and this will have a devastating effect on life, plants and animals. The sun will be blocked for months by this dust; the soil will be polluted just like in a volcanic eruption, ashes fall all over the planet. Imagine the worst wildfire you have ever seen on the news, now think of one that covers South Africa, Namibia, Angola, Botswana, Madagascar, Zimbabwe, Mozambique, Malawi, Zambia, Tanzania, and the Congo, almost a third of Africa. A mega blaze, meanwhile the molten rock droplets are being carried around the globe, this material heated up the atmosphere up to temperatures similar to those of an impact, evaporating sea water, and on land causing plants to burst into flames. The soot of these fires was raised to the upper atmosphere by hot air and then it fell over the already hot material ejected by the impact, devastating the environment even more.

2,000 to 20,000 mile and hour Winds are coming; they will arrive in India in less than 2 hours.

But the worst is yet to come, an impact this size created a massive void in the sea and the water is rushing back to fill it, this is creating a wave the size of which has not been seen for millions of years, a mega tsunami is in the making, people were far away and didn't know it was happening, but their fate had been sealed.

The wall of water rose, and kept rising until it reached 200 to 210 meters high (650 feet), this was not a coastal wave, it did not roll and break, this was a tsunami, which is a mountain of water, and let's remember that in open ocean, tsunamis travel at very high speeds and lose very little or no energy at all in the process, a speed between 700 and 800 kilometers per hour (500 miles per hour) is calculated for a Tsunami in open seas. While this wave traveled, the wave of a mega earthquake traveled through the sea floor.

Meanwhile, 3 miles underwater, the Earth's crust has been molten by the impact. As a reference, let me tell you about large tsunamis that occurred just a few decades ago.

In 1960, tsunami waves caused by the 9.5 magnitude Valdivia earthquake in Chile (the most powerful in recorded history) reached Japan, more than 16,800 km (10.600 miles) away in less than 24 hours, killing hundreds of people.

For reference, the last large tsunami in Polynesia in 2009 recorded wave heights of 14 meters (46 ft) at their highest on the Samoan coast, this one was 210 meters (650 feet) high.

So, Noah and other people like him are about to be hit, first by winds of force never felt by humans before and then by a 200 meter wave, the coast of Madagascar is within the kill zone, it gets hit by winds with a speed of thousands of miles an hour, carrying water, dust, molten rock, pebbles and the topsoil of the island is sandblasted away; the eastern shore of Africa is a bit more protected by Madagascar itself but fares no better, as I already mentioned, no one survived in that area to tell us what happened, so no flood myths there, we have to look at the geology of the place to find out. As a reference, Buckshot travels at about 800 mph when fired, so you can imagine that some places were hit by winds carrying gravel, rocks sand and red hot material. Impossible to survive something like that being that close.

Going back to the myths, around the Indian Ocean, islands such as Pulau Nias and Palau Engano, west of Sumatra, tell of people saved by climbing on mountain peaks rising between 200 to 300 m above sea level, there is a Tamil story from India that tells us that "the sea rose flooding the city of Maturai," reaching a level between 100 and 200m above sea level and almost 100 km inland from the coast.

So, a wave, 200 meters (650 feet) high is racing unchecked away from Burckle crater, in all directions at a speed between 700 and 800 kilometers per hour (500 miles per hour), and will not lose energy until it reaches destination, however, according to the scientific paper mentioned above, the comet might have been a Shoemaker-Levy type body, broken up by gravity in a previous orbit around the sun and causing two or more other major impacts and several minor ones at the same time, the craters of these other impacts have not been found yet, or may I say, some impacts have been found, but not yet linked to the Burckle impact, but modeling based on the global distribution of flood myths places one of them close to Easter island in the mid south pacific and the other close to the Aleutian islands in the Bering Sea.

I sincerely hope researchers find these other craters (if they exist) and we can plot a more precise picture of the events.

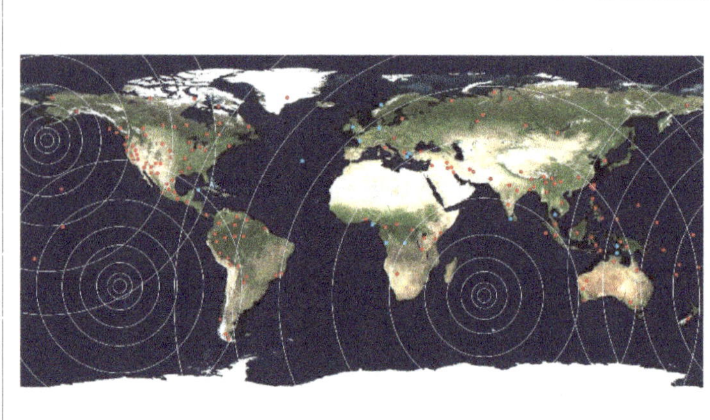

But for now we will focus on the one in the Indian Ocean, it has been properly dated to the time of Noah based on 2 drill cores, and the impact; although insufficient to cause a "global deluge" by itself, it most certainly was powerful enough to cause a catastrophe to remember as God's wrath for generations.

Now, in the standard story, we are told Noah is chosen by god to "restart" humanity because he was "...a righteous man and blameless in his time, but Noah was not just any human, he was the grandson of Enoch; in the Bible, Noah, was said to be the son of Lamech and his wife Betenos; however, when the dead sea scrolls were discovered, and more specifically the book of Enoch; religious scholars were baffled by a very different version of events, an earlier version that told a different and earlier story. Lamech's parchment tells the story of Noah, and that he had been chosen by god even before he was born; so the story says that when Noah was born, his skin was whiter than that of his brothers, his hair was also white and in the words of Lamech "This child can't be my son, he looks like the guardians, when he opens his eyes all the house is illuminated". Lamech was away when Noah was born, he returned home after the birth, and he had been away less than nine months, so Lamech seeks counsel from his father Methuselah, who in turn seeks counsel form his father Enoch who says to him "The guardians from heaven impregnated Betenos without touching her, Lamech must accept this child as his own and name him Noah, because the guardians have decided that Noah will be the father of the future generations."

In the Bible, it is God who instructs Noah on how to build the ark: "…This is how you are to make it: the length of the ark 300 cubits, its breadth 50 cubits, and its height 30 cubits," roughly two thirds of the Titanic.

And: "…Each deck has to be the same height as the Temple in Jerusalem, and each has to be the triple of the area of the court of the tabernacle…" but in the book of Enoch, it is not God who tells Noah to build the ark, it is the Archangel Uriel. It is Uriel who warns Noah about the coming deluge and tells him about the future of humanity after the cataclysm.

Some religious scholars and scientists have told us that this event is an allegory, that it is not to be taken literally; but after comparing it to hundreds of flood myths, and now having "a messenger" instead of God himself, the story matches the "witness accounts" even better.

Now, "Around 2300 B.C., or about 4,300 years ago plus or minus 100 years, many of the major civilizations of the world collapsed. The Akkadian Empire in Mesopotamia, the Old Kingdom in Egypt, the Early Bronze Age civilization in Israel, Anatolia and Greece, as well as the Indus Valley civilization in India, the Hilmand civilization in Afghanistan and the Hongshan Culture in China - the first urban civilization in the world - all collapsed at more or less the same time."

From: "Comets and Disaster in the Bronze Age", Benny Peiser

Spread of the Mega-tsunami and shockwave of the impact

As an example, I read a theory on the collapse of the Akkadian Empire, those who advocate for the theory, stated that the soil got more saline over the years due to salt content in the irrigation water and that crops wouldn't grow anymore. However, although this might possibly be true as an explanation, we have to keep in mind that we know for a fact that a massive tsunami generated by the Burckle impact hit the coast of the Akkadian empire and that the wave is recorded reaching up more than 100 miles inland in India, if you look at this map, we can be almost certain that agricultural land in Akkad was flooded with sea water.

Let's keep in mind that Fallujah is 43 m.a.s.l and Baghdad is 34 m.a.s.l. and they are several miles upstream of where Akkad was located. So speaking in meters, we calculated the tsunami at about 200 meters high, so the wave should have ran up the Tigris and Euphrates unchecked for tens of miles. If it wasn't the wave, witness accounts tell us of fire particles and embers raining in India, and this leads us to assume the deposition of minerals from a dust cloud caused by the impact might have altered the soil as well. Also, we have to remember that in India they saw the following "the sea rose flooding the city of Maturai," reaching a level between 100 and 200m above sea level and almost 100 km inland from the coast. 100 kilometers inland, and Maturai was 4 times higher above sea level than Akkad.

Another example of possible evidence of major environmental damage to the region is further North, to be more precise in Turkey. This is still a bit controversial, but under the city of Derinkuyu in 1962, a farmer was renovating his basement and found a room, this room ended up being part of a whole city which seems to have housed 20,000 people, the construction reaching 85 meters underground in 11 levels, with churches, farms, wineries and which had an extremely complex ventilation for when it was closed, with circular stone doors which were rolled into place.

This city is said to be 2,700 years old by the Turkish government, but many researchers believe it is about 4,000 years old. Why would people go and live underground? Some archaeologists say it was to protect themselves from invaders and their raids, but this doesn´t make much sense, if you lock yourself underground, all the invaders have to do is find your ventilation shafts and block them or light fires in them and "smoke" you out, your shelter then becomes your death trap. Many "fringe" researchers believe that when Derinkuyu and many other underground structures were originally planned and built, the purpose was to be able to survive underground because the surface was uninhabitable or at least a very hostile environment. If these cities were "forts", why would you have farms inside? Massive storage facilities would have sufficed, if the stores of food were large enough, people would have been able to live underground for months at a time. Now, the reason to have underground farms could also be that the surface was not healthy to live on or to grow things on; yes, they could have been able to go out for short periods of time, maybe even a day or two, but maybe animals couldn't graze, or the dust might have been toxic if they were exposed to it for too long. Yes, these cities were later used as forts, but they don´t seem to have been designed to be used for war shelters, they seem to be designed almost as survival arks. Only time and research will tell if these underground cities were used to survive Noah's flood, or, to be more specific, the aftermath of an impact of planetary consequences.

Photo: Depositphotos.com

There is one place where no one has looked yet, it's Antarctica, just like there was a Greenland ice core project, an Antarctica ice core project could shed some light by identifying whatever material was hurled into the atmosphere by this impact, and also maybe have recorded the impact of the tsunami in sediments and other indicators that might have been preserved in the environment of the least disturbed place in the planet.

Now, what about the future?

We know it has happened before, Earth has been hit by different types of external forces, coronal mass ejections, EM pulses, asteroids, comets, you name it; occurrences like the Carrington Event, which was a geomagnetic storm caused by a coronal mass ejection that happened on September 1st and 2nd 1859, actually Earth and Mankind were lucky that it didn't happen today, the estimate damage of such an event today is one to two trillion dollars. The estimate periodicity of such events is 150 years, so we are due for another pretty soon; the electromagnetic charge of the Carrington event (named after an amateur astronomer named Richard Carrington), charged telegraph wires, electric arcs were reported shocking telegraph operators and telegraph stations caught fire, the operators notice they could disconnect the batteries and send messages o from Boston to Portland, Maine using the current in the atmosphere. This was said by a woman in Charleston. "The eastern sky appeared of a blood red color. It seemed brightest exactly in the east, as though the full moon, or rather the sun, were about to rise. It extended almost to the zenith. The whole island was illuminated. The sea reflected the phenomenon, and no one could look at it without thinking of the passage in the Bible which says, 'the sea was turned to blood.' The shells on the beach, reflecting light, resembled coals of fire." Today it would certainly damage the internet, communications, satellites, it would be a catastrophe. Well, this is just energy, not an impact

We have to understand that the K-T event that killed off the dinosaurs (along with 75% of all life on the planet) hit Earth hard, the recovery was long, scientists estimate it took Earth 4 million years to recover the ecosystem, it took that long to recover biodiversity in South America, and yet once earth recovered, many species simply never recovered, all non avian dinosaurs became a memory, along with thousands of species of plants and insects. If you think 4 million years is a long time, think again, scientists estimate it took 9 million years for North America to Recover.

So, is Earth like it's supposed to be today, or are we still recovering from the impact that is believed to have ended the last ice age 12.000 years ago, or from the Burckle impact 4.000 years ago? We really don´t know, what we do know is that these things have happened before and will happen again. This is not me telling you to go and build a bomb shelter or to buy land in a valley in the Himalaya or in the Rocky Mountains, I am just asking you to think differently about our planet, it is not the constant benign environment that many believe; it is a dangerous living planet, it is a dangerous living solar system, climate is not stable, it has been for the last 10 or 12.000 years, but that is not the history of our climate. Would our climate be stable if it wasn't altered by cosmic events? Maybe, but please keep in mind that these impacts have happened in the past and had extreme effects on life on Earth. I hope humanity doesn't need to "start again like children" in the near future, but the fact that we don´t want it to happen doesn't really matter, it will happen; the only question is when.

Conclusion on Noah's Flood

So, this is the information that has pointed me into believing Noah's flood was an impact event about 4,000 years ago, if I am right or wrong it is for you to decide. This comes from different historical, religious and scientific sources. There is tons of information out there for you to research, and I didn't want to include it all here because a 1,000 page book was not the point. If you read the Bible and for you it is a record of real events, I hope I helped steer you to information on events that can be pinpointed in the Bible, events that are scientifically proven and that provide evidence for events that have been known to hundreds of cultures for millennia, and also to make those who don´t believe in the historical value of the Bible, the Khuran and other books, aware that religious texts and traditions are not just about religion. On the other hand, if you are a person who has never read religious texts and has always regarded religion as myth, I hope I have opened your mind just a little bit so you can look at the Bible or other religious texts and traditions in a different way, and that maybe you will see the truth, not between the lines, but plain and clear in those lines that you previously disregarded. Should we read the Book of revelation to try to find out how the world might end? It's actually your choice, we know that those who listened once (like Noah or Utnapishtim) were saved, now maybe time for you to read the book of revelation and maybe what it says will not sound so farfetched as before.

www.ingramcontent.com/pod-product-compliance
Lightning Source LLC
Chambersburg PA
CBHW071923120726
48001CB00005B/1846